The Midjourney Ex[

Generate creative images from text prompts and seamlessly integrate them into your workflow

Margarida Barreto

<packt>

The Midjourney Expedition

Copyright © 2024 Packt Publishing

All rights reserved. No part of this book may be reproduced, stored in a retrieval system, or transmitted in any form or by any means, without the prior written permission of the publisher, except in the case of brief quotations embedded in critical articles or reviews.

The author acknowledges the use of cutting-edge AI, such as ChatGPT, with the sole aim of enhancing the language and clarity within the book, thereby ensuring a smooth reading experience for readers. It's important to note that the content itself has been crafted by the author and edited by a professional publishing team.

Every effort has been made in the preparation of this book to ensure the accuracy of the information presented. However, the information contained in this book is sold without warranty, either express or implied. Neither the author, nor Packt Publishing or its dealers and distributors, will be held liable for any damages caused or alleged to have been caused directly or indirectly by this book.

Packt Publishing has endeavored to provide trademark information about all of the companies and products mentioned in this book by the appropriate use of capitals. However, Packt Publishing cannot guarantee the accuracy of this information.

Group Product Manager: Niranjan Naikwadi
Publishing Product Manager: Tejashwini R
Senior Editor: Mark D'Souza
Technical Editor: Simran Ali
Copy Editor: Safis Editing
Proofreader: Mark D'Souza
Indexer: Manju Arasan
Production Designer: Ponraj Dhandapani
DevRel Marketing Coordinator: Vinishka Kalra

First published: April 2024

Production reference: 1170424

Published by Packt Publishing Ltd.
Grosvenor House
11 St Paul's Square
Birmingham
B3 1RB, UK.

ISBN 978-1-83508-697-1

www.packtpub.com

To my adorable Leia and Luke

– Margarida Barreto

Contributors

About the author

Margarida Barreto blends over 20 years of expertise in communication design with a fervent passion for AI, crafting innovative campaigns that captivate and inform. Her journey spans graphic and web design, marketing, and social media, collaborating with notable clients such as Apple, HP, Dell, and Mitsubishi. Recently, Margarida has focused on merging visual communication with AI, particularly in technological and lifestyle brands, to create unique experiences that resonate across digital landscapes. As a creative director, she excels in team leadership, emphasizing the synergy of collective creativity and technological innovation to push the boundaries of art and design.

I extend my heartfelt thanks to my husband, Gonçalo, for his unwavering support on my AI journey, and to the entire AICC community on LinkedIn, Discord, and Instagram. Your incredible support and creativity have been a constant source of inspiration and encouragement.

About the reviewers

Lakshmi Narayanan V is an architect from India. He works at an architecture firm based in New Delhi. His interest in computation and advanced architectural design has led him to take part in various events outside of work, such as the Structural Integrity Summer School at KU Leuven. He is also involved in computational research work and has delved into spatial data science using ArcGIS. His quest to learn and explore technological advancements introduced him to Midjourney, Grasshopper, and generative design. He is also a keen reader and believes in journalism as a tool to bridge the gap between people and architects. Other than his interest in architecture, he is an adventurer who travels to different places every three months, exploring the nature of the Earth. His other hobbies are early morning runs, playing the violin, food, and watching movies.

Brian Hmurovich is the author of *Goose City* on Kindle and the owner and operator of Aifotostock.com, a print-on-demand website using AI images. He has been a Midjourney user since August 2022. Previously, he was a professional event photographer in San Diego, CA, for five years, working for companies such as Petco Park, Red Bull, Be Water Photo, Network After Work, Nite Guide Magazine, LED, NBC San Diego, and many local small businesses.

Most importantly, he is a loving, dedicated family man who loves to go on hikes, enjoys walking his chihuahuas, Odie and Luna, going to drive-ins, traveling the world, and, of course, playing video games with his daughter.

Table of Contents

Preface — xi

Part 1: Getting Started and Exploring Midjourney

1

Exploring the Midjourney AI World — 3

A brief introduction to the world of AI	4	What is Midjourney?	9
The birth of AI	5	Understanding Midjourney AI's technology	9
LLMs in AI	6	The versions' evolution	10
Ethical considerations in AI	6	Legal challenges and ethical concerns around Midjourney	10
Merging AI with art	6	Exploring the benefits of Midjourney	12
Applying AI in generative art	6	Summary	15
What the future holds for AI art	9		

2

Embarking on a Journey – From Discord to Midjourney — 17

How to join the adventure	18	Creating a server	27
Diving into Discord	18	Inviting the Midjourney Bot to your server	28
Logging in to Midjourney and connecting to Discord	21	Start prompting	29
Want quieter? Inviting the Midjourney Bot to your server	27	Summary	30

Part 2: Unlocking the Power of Midjourney – A Deep Dive into Features and Functionalities

3

Mastering Midjourney Versions – Quick Start Guide   35

Technical requirements	36	Midjourney's evolution	49
Imagine and prompt	36	V1 and V2	49
My first prompt – now what?	39	V3 and V4	54
Beyond /imagine – command list	45	V5 to V5.2 (the current model)	60
Taking a journey through		Niji versions – the anime world	64
		Summary	69

4

Understanding and Learning Parameters   71

Technical requirements	72	Basic parameters	74
What are parameters?	72	Legacy and special parameters	97
List of parameters and examples	74	Summary	101

Part 3: Advanced Prompting and Visual Creations

5

Navigating through Advanced Prompts   105

Technical requirements	105	Styles and aesthetics	126
Blend mode and image prompting	106	Nested permutations	128
Blend mode	106	Weights and image references	130
Image prompting	109	Summary	131
Multi-prompting	113		
Permutations	118		
Permutations and parameters	119		

6

Upgrading Your Prompt for Optimal Results 133

Technical requirements	134	Photographic styles	158
Help me describe	134	Light	165
A world of styles	140	Camera angles	170
Art styles	143	**The right words**	**175**
Mood styles	157	**Summary**	**183**

7

Customizing Midjourney – Settings, Preferences, and Unleashing Creative Prompts 185

Technical requirements	186	Fine-tuning our images with the Style Tuner	200
An overview of Midjourney's settings	186	Combining style codes	205
Customizing your preferences	196	Adjusting with --stylize and --raw	206
/prefer option	196	Experimenting with random styles	208
/prefer suffix	198	**Summary**	**212**

Part 4: Prompting for the Real World

8

Exploring Practical Use Cases and Pushing Boundaries 215

Technical requirements	216	The visual power of storytelling	223
Generating ideas and captivating moodboards	216	Practical example	224
Business and marketing strategies	218	Creating brand sets with Midjourney – icons and logos	230
Event planning and celebrations	219	Understanding the essence of brand identity	230
Interior design projects	220	Leveraging Midjourney's features for precise control	233
Conceptualizing brand identities	221		
Creative brainstorming	222		

Utilizing Midjourney's photorealism for product mockups	239	Creating professional product mockups with Midjourney	245
Practical applications	241	Summary	252

9

Unlocking Tips and Tricks and Special Functions — 253

Technical requirements	254	Exploring new upscaling options in Midjourney V6 ALPHA	265
Using the face you want with InsightFaceSwap Bot	254	Using third-party upscaling tools	268
Creating high-resolution imagery for printing	262	Tips and tricks to improve your prompts	272
Introducing Midjourney's new upscalers – Upscale 2x and Upscale 4x	263	Summary	282

10

Conclusion: Your Journey with Midjourney — 285

Reflecting on the journey and its potential	286	The evolution of content creation into a neural future	292
My point of view and perspectives – the next leap, AI's role in our creative and everyday lives	288	Ethical landscape and the path forward	294
		Summary	296

Index — 299

Other Books You May Enjoy — 306

Preface

This book embarks on an exploratory journey into the fascinating world of **artificial intelligence** (**AI**) and generative art, with a special focus on the revolutionary tool known as Midjourney. Throughout this book, we explore the fundamentals of AI and generative art, looking into advanced techniques and practical applications of Midjourney. This comprehensive guide is designed for artists, designers, and technology enthusiasts eager to explore new frontiers of AI-assisted creativity, offering deep insights, advanced techniques, and a broad overview of the possibilities that this emerging technology presents to the creative and artistic fields.

The increasing importance of Midjourney and AI-assisted generative art cannot be understated. We are witnessing a transformation in how artists, designers, marketing professionals, architects, and many others in creative fields are creating and showcasing their work. This book highlights how Midjourney is redefining the boundaries of creativity, enabling these professionals to transcend traditional barriers of artistic expression and forge new paths in their respective fields.

Who this book is for

This book is designed for a broad spectrum of readers – from seasoned artists and designers seeking to integrate AI into their repertoire to novices curious about the intersection of technology and art. It is particularly valuable for those in creative industries who are constantly seeking innovative ways to communicate visually and differentiate their work in a competitive landscape.

What this book covers

Chapter 1, *Exploring the Midjourney AI World*, serves as an introduction to the fascinating world of AI and generative art, with a special focus on the Midjourney tool. Here, you will discover how AI is transforming artistic creation, offering an overview of fundamental concepts and the limitless possibilities that Midjourney unlocks for creatives across all fields.

Chapter 2, *Embarking on a Journey – From Discord to Midjourney*, guides you in setting up and getting started with Midjourney through Discord, from creating an account to executing your first prompts. This chapter provides a step-by-step guide for beginners, ensuring a smooth transition into the world of AI-assisted generative art.

Chapter 3, *Mastering Midjourney Versions – Quick Start Guide*, takes you through the different model versions of Midjourney and how each can be utilized to achieve specific results. This chapter offers insights into selecting the appropriate version for your projects, maximizing the quality and precision of the generated images.

Chapter 4, *Understanding and Learning Parameters*, shows how you can customize your creations in detail. This chapter explores the importance of each parameter and how they influence the final outcome, allowing for more refined control over the creative process.

Chapter 5, *Navigating through Advanced Prompts*, helps you advance your prompt creation skills with advanced techniques and strategies to maximize the effectiveness of your generated images. This chapter addresses the art of constructing complex prompts, using images and multi-prompts to inspire unique and detailed creations.

Chapter 6, *Upgrading Your Prompt for Optimal Results*, helps improve your prompt skills with advanced techniques and additional features. This chapter is dedicated to enhancing the precision and detail of your prompts, introducing resources such as the *describe* command to enrich your artistic vocabulary and explore varied styles for extraordinary results.

Chapter 7, *Customizing Midjourney – Settings, Preferences, and Unleashing Creative Prompts*, shows how you can personalize your Midjourney experience by adjusting settings and preferences to meet your creative needs. This chapter guides you through customizing Midjourney, from using Midjourney's Style Tuner to creating custom codes for frequently used values in your prompts.

Chapter 8, *Exploring Practical Use Cases and Pushing Boundaries*, covers how Midjourney can be adapted to meet professional needs across various fields. This chapter presents practical applications of Midjourney, from creating visual narratives to developing branding and product design, demonstrating the tool's versatility in different creative contexts.

Chapter 9, *Unlocking Tips and Tricks and Special Functions*, uncovers advanced techniques and special functions of Midjourney that can transform your creative process. This chapter unveils valuable tips and tricks for optimizing the use of Midjourney, in addition to introducing special functionalities that expand the possibilities of creation.

Chapter 10, *Conclusion – Your Journey with Midjourney*, reflects on the transformative journey with Midjourney and contemplates the future potential of AI in art and design.

To get the most out of this book

It is recommended that you have access to a computer with an internet connection and are willing to actively explore the exercises and practical examples presented. A basic understanding of technology and web navigation is beneficial, but the book's clear, step-by-step guidance makes it accessible to individuals at all skill levels.

Software/hardware covered in the book	Operating system requirements
Discord	Windows, macOS, or Linux
Midjourney	Windows, macOS, or Linux
ChatGPT	Windows, macOS, or Linux

To get the most out of this book, you are recommended to have a basic understanding of technology concepts; it would be a plus to be familiar with the technical language common in the fields of creativity, marketing, design, architecture, and animation. This familiarity will be significant when writing prompts and providing guidelines to the AI, allowing for a deeper and more effective exploration of Midjourney's potential.

Conventions used

There are a number of text conventions used throughout this book.

`Code in text`: Indicates code words in text, database table names, folder names, filenames, file extensions, pathnames, dummy URLs, user input, and Twitter handles. Here is an example: "When using images in your prompts, ensure that the image address is a direct link to an online image, ending with extensions such as `.png`, `.gif`, `.webp`, `.jpg`, or `.jpeg`."

Bold: Indicates a new term, an important word, command names, or words that you see onscreen. For instance, words in menus or dialog boxes appear in **bold**. Here is an example: "This will open a page with an invite to join the Midjourney Discord channel. Click **Accept Invite**. Enter a username of your choice and select **Continue**."

> **Tips or important notes**
> Appear like this.

Get in touch

Feedback from our readers is always welcome.

General feedback: If you have questions about any aspect of this book, email us at `customercare@packtpub.com` and mention the book title in the subject of your message.

Errata: Although we have taken every care to ensure the accuracy of our content, mistakes do happen. If you have found a mistake in this book, we would be grateful if you would report this to us. Please visit `www.packtpub.com/support/errata` and fill in the form.

Piracy: If you come across any illegal copies of our works in any form on the internet, we would be grateful if you would provide us with the location address or website name. Please contact us at `copyright@packt.com` with a link to the material.

If you are interested in becoming an author: If there is a topic that you have expertise in and you are interested in either writing or contributing to a book, please visit `authors.packtpub.com`.

Share Your Thoughts

Once you've read *The Midjourney Expedition*, we'd love to hear your thoughts! Scan the QR code below to go straight to the Amazon review page for this book and share your feedback.

https://packt.link/r/1-835-08697-7

Your review is important to us and the tech community and will help us make sure we're delivering excellent quality content.

Download a free PDF copy of this book

Thanks for purchasing this book!

Do you like to read on the go but are unable to carry your print books everywhere?

Is your eBook purchase not compatible with the device of your choice?

Don't worry, now with every Packt book you get a DRM-free PDF version of that book at no cost.

Read anywhere, any place, on any device. Search, copy, and paste code from your favorite technical books directly into your application.

The perks don't stop there, you can get exclusive access to discounts, newsletters, and great free content in your inbox daily

Follow these simple steps to get the benefits:

1. Scan the QR code or visit the link below

   https://packt.link/free-ebook/9781835086971

2. Submit your proof of purchase
3. That's it! We'll send your free PDF and other benefits to your email directly

Part 1: Getting Started and Exploring Midjourney

This part lays the foundational knowledge necessary for readers to begin their journey into artificial intelligence (AI)-assisted creative processes. This section not only introduces the revolutionary tool Midjourney but also demystifies the core concepts of AI and generative art. As the digital landscape evolves, understanding these fundamentals is crucial for any creative professional looking to leverage AI to enhance their design workflows and visual communication skills.

This part has the following chapters:

- *Chapter 1, Exploring the Midjourney AI World*
- *Chapter 2, Embarking on a Journey – From Discord to Midjourney*

1
Exploring the Midjourney AI World

Artificial intelligence (**AI**) and **generative art** have forged an unexpected alliance to produce unique outcomes in the world of design and aesthetics. Today, an impressive tool known as **Midjourney** stands at the center of this fusion, harnessing the power of AI to produce stunning visual results with unparalleled ease.

This chapter takes you on a journey through the genesis and realm of AI, introducing generative art and its exciting application in Midjourney. It concludes with a dive into real-world examples that highlight the utility and potential of AI-powered creativity. Whether you're a marketer, designer, or enthusiast keen on amplifying your creative output, understanding the foundations and applications of Midjourney is key to unlocking your AI-assisted creative potential.

In this chapter, we're going to cover the following topics:

- **A brief introduction to the world of AI**: Here, we will set the stage for understanding how AI influences numerous aspects of modern life and creative processes. We will explore the fundamentals of AI and discuss its importance, evolution, and the various subfields that compose this groundbreaking technology.

- **Merging AI with art**: Here, we will explore how AI is being used to create new forms of generative art, transforming traditional notions of creativity and artistic expression.

- **What is Midjourney?**: Here, you will find a comprehensive overview of Midjourney, detailing its role as a prominent AI tool in the universe of generative art and why it stands out in the crowded space of **generative AI** (**GenAI**).

- **Exploring the benefits of Midjourney**: Discover the vast benefits and practical applications of using Midjourney in various creative endeavors. From boosting efficiency and fostering innovation to its impact on marketing and beyond, you can see the potential of Midjourney across different industries.

By the end of this chapter, you will have a holistic understanding of AI, its intersection with generative art, the key role Midjourney plays in this confluence, and the tremendous benefits it offers for art creation. Not only will you learn about these technologies theoretically, but you'll also gain an understanding of their practical implications and benefits.

A brief introduction to the world of AI

AI is a rapidly evolving field that is transforming our world in numerous ways. Today, AI can be found powering virtual assistants, recommending movies on streaming platforms, diagnosing diseases in healthcare, driving autonomous vehicles, and even assisting climate change research. This revolutionary technology, deeply rooted in disciplines such as computer science, mathematics, cognitive psychology, and philosophy, aims to construct machines and systems capable of performing tasks that usually require human intelligence.

Notably, AI isn't a **monolithic domain**. It comprises various subfields, each focusing on distinct aspects of AI.

> **What does *monolithic* mean?**
>
> The term *monolithic* often refers to something that is large, powerful, and intractably indivisible or uniform. In the context of the sentence "*AI isn't a monolithic domain*," it means that AI isn't just one big, unified field, but rather consists of various diverse subfields, each with its unique focus and specialization.

Here are some key subfields of AI:

- **Machine learning**: Used by Midjourney, this subset of AI empowers machines to learn from data without explicit programming. Machine learning algorithms are trained on extensive datasets; they predict, classify, or make decisions based on this training.
- **Natural language processing (NLP)**: NLP facilitates the interaction between computers and human language. It allows computers to understand, interpret, and generate human language in a valuable way.
- **Computer vision**: This subfield, essential to Midjourney, involves teaching computers to interpret visual data. Computer vision algorithms extract insights from images and videos, identifying objects, tracking movements, or recognizing patterns.

To better understand the potential of AI, let's explore its different categories:

- **Narrow AI**: Systems designed to perform a specific task, such as recommending songs, answering questions, or predicting the weather. Midjourney falls into this category, as it is designed to perform the specific task of converting text prompts into images.

- **General AI**: Systems capable of understanding, learning, adapting, and implementing knowledge in a way that can effectively substitute a human. Midjourney does not fall into this category, as it's a specialized system rather than a generalized one.
- **Superintelligent AI**: An intellect that outperforms the best human brains in every field, including scientific creativity, general wisdom, and social skills. Midjourney's functionality doesn't reach this advanced level of intelligence.

Having explored the different categories of AI, we turn now to the origins of AI, tracing back the concept's roots and evolution. Understanding this history provides a foundation to appreciate the complex trajectory of AI development, leading us to a closer examination of **large language models** (**LLMs**) and ethical considerations that have arisen in the AI era. From the birth of AI to the cutting-edge innovations and debates of our time, we'll explore milestones and challenges that have shaped this fascinating field.

The birth of AI

The concept of AI is not new; the idea of crafting machines capable of reasoning like humans traces back to ancient times. However, the birth of modern AI as we understand it today commenced at a seminal conference at Dartmouth College in 1956. It was here that the term *artificial intelligence* was coined and became the banner under which a new era of technological discovery would march.

At this landmark event, leaders in the field converged on a hypothesis that *"every aspect of learning or any other feature of intelligence can, in principle, be so precisely described that a machine can be made to simulate it."* Figures such as Alan Turing, celebrated for the Turing Test, and John McCarthy, often acknowledged as the father of AI, played instrumental roles in these formative stages.

Since then, the development of AI has seen impressive highs and discouraging lows, culminating in the extraordinary advancements we observe in today's world, such as LLMs.

Figure 1.1 – Can machines reason like humans? (Created with Midjourney by the author)

LLMs in AI

LLMs are a significant development in AI. These models generate human-like text, having been trained on extensive text data. GPT-4 by OpenAI, Jurassic-1 Jumbo by OpenAI, and Megatron-Turing NLG by Nvidia are prominent examples of LLMs. These LLMs find applications in areas such as realistic dialogue creation for chatbots, creative text generation for ads or scripts, and informative response delivery to user queries in various contexts. Despite their impressive capabilities, they do have limitations, as they don't truly comprehend the text they generate and can sometimes produce misleading or biased outputs.

Another crucial aspect to consider when using AI, especially given the capabilities and limitations of these large models, is ethics.

Ethical considerations in AI

As AI continues to advance, ethical considerations have become increasingly important. AI systems can inadvertently perpetuate existing biases that can result in discriminatory outcomes. For instance, a hiring algorithm trained on a dataset of resumes from a company with a history of gender bias might discriminate against female applicants. Similarly, AI technologies such as facial recognition can invade people's privacy and be misused in surveillance.

Addressing these concerns and creating ethical, fair, and transparent AI systems is critical as we move forward in the AI era.

With a foundation laid in understanding AI, its development, limitations, and ethical considerations, we are poised to explore an exciting frontier where AI transcends traditional boundaries: the world of generative art. This next section unveils how algorithms and creativity merge, forging a new genre that challenges our perceptions of art and creation.

Merging AI with art

As we venture deeper into the intersection of technology and creativity, we find that AI has innovatively penetrated the realm of art, surprising many of us and giving birth to a fascinating new genre: generative art. Blending the calculated precision of algorithms and the whimsical element of randomness, AI has revolutionized generative art, churning out mesmerizing and distinctive pieces.

Applying AI in generative art

Generative art refers to any art practice where the artist leverages an autonomous system to contribute to or decide upon the final outcome, and it can be music, an image, or even a video. These systems are based on machine learning powered by algorithms, mathematical functions, and data, and can mimic human actions such as generating art. A defining characteristic of generative art is the shift in the artist's role. In traditional art, the artist is the only creator of the work. They have complete control over the creative process, from concept to execution. However, in generative art, the artist shares the control with the system or algorithm. The artist still has influence, but now they are not the only one creating it. This shift has various implications. First, it means that generative art can be created by anyone, opening this field of art to a wider range of people. Second, it can create art much more quickly than traditional art and also be more accessible and affordable than traditional art forms. And lastly, it can be more complex and unpredictable as the artist has less control over the creative process and, in the end, depends on what the system will generate.

AI, with its profound capacity to recognize patterns, decipher complexities, and generate new outputs, brings an unprecedented dimension to generative art. Machine learning models, including those based on **neural networks** (**NNs**), such as **generative adversarial networks** (**GANs**) and **variational autoencoders** (**VAEs**), as well as **diffusion models** (though these models can involve NNs in their implementation, diffusion models aren't strictly classified as NNs), have exhibited extraordinary proficiency in learning and replicating artistic styles to create original pieces.

> Neural networks
>
> NNs are computational models inspired by the way biological NNs in the human brain work. They consist of layers of nodes, or "neurons," that can adapt to patterns in data.

A notable example of NN-based generative art is *Portrait of Edmond de Belamy*, created by a GAN, which fetched an astonishing $432,500 at Christie's auction house.

GANs, VAEs, and diffusion models have been particularly influential in the field of AI art. GANs comprise two parts – a generator that creates new data instances, and a discriminator (a classifier that determines if the input samples are real or fake) that assesses them for authenticity. VAEs excel in generating new samples in the latent space. Diffusion models frame data generation as a reverse diffusion process. This means that these models gradually refine a simple noise input and progressively refine the input to generate data similar to the data on which the model was trained (*Figure 1.2*). This attribute makes them especially useful in generating high-resolution images.

Figure 1.2 – Diffusion models can be used to generate images from noise (Created with Midjourney by the author)

> **Latent space**
>
> The latent space in a machine learning model represents a set of variables that influence specific characteristics of a data distribution. For instance, in a dataset of cars, the latent space might include variables such as color, orientation, or the number of doors. However, defining the role of each component in the latent space becomes complex, especially when dealing with high dimensions. There may also be dependencies between components, further complicating the manual design of this space. Thus, defining this complex distribution P(z) proves challenging.

Within the domain of art, these algorithms are trained to generate novel pieces (*Figure 1.3*) while ensuring they align with the artistic style or aesthetic they have been trained on:

Figure 1.3 – The Mona Lisa if it were created in pop art style (Created with Midjourney by the author)

The outcome is an AI system that can produce art, frequently blurring the line between human and machine creation. This progressive technology prompts us to reconsider our definitions of art, the role of the artist, and the value of creativity.

What the future holds for AI art

The trajectory of AI in art presents a plethora of opportunities and challenges. On one hand, there's the tantalizing possibility of AI art reaching a level of sophistication that renders it indistinguishable from human-created art. This evolution could provoke intense debates on authenticity, originality, and the valuation of AI-created art. On the other hand, AI could become a remarkable tool for fostering new and innovative forms of expression, broadening our understanding of art itself.

However, this significant progress in AI art is not without its ethical conundrums. As AI-generated art becomes more prevalent, a primary concern is its potential misuse, such as creating deepfakes or other forms of synthetic media that could be used to deceive or manipulate. These developments necessitate a critical appraisal of ethical boundaries in AI art.

As AI and generative art continue to intertwine, the roles of the artist, the viewer, and the AI itself in the creative process become subjects of redefinition. This fascinating interplay promises a dynamic future for the world of art, one where creativity and technology unite to push the boundaries of our imagination.

Having examined the complex interplay between AI and generative art, with its fascinating potential and ethical dilemmas, we now turn our attention to a specific application that is the focus of our book: Midjourney. In the upcoming section, we will uncover its origin, functionality, technological underpinnings, and the broader ramifications it has for the realm of art and creativity.

What is Midjourney?

Midjourney is a powerful AI text-to-image generation tool that serves as a bridge between AI and generative art. It leverages advanced AI algorithms to help users create visually captivating designs and artwork. Its sophisticated AI engine generates art based on user input, making it an ideal tool for anyone seeking to incorporate unique visuals into their work. This innovation has garnered significant attention and even stirred debates about the necessity of human creativity in the future. But is it truly the end for artists? To answer this question, it's crucial to understand the inner workings, capabilities, and limitations of Midjourney.

Understanding Midjourney AI's technology

Midjourney is a product of a self-funded independent research institute, led by David Holz (co-founder of Leap Motion); it was conceived to generate images from text. On July 12, 2022, the Midjourney team rolled out the beta version.

Similar to other projects such as DALL-E from OpenAI and Stable Diffusion, Midjourney distinguishes itself through its performance and the quality of images it can generate.

Users can currently create high-quality images through Discord bot commands, with no special hardware or software required. The main command is **/imagine**, which prompts the bot to generate an image. However, plans for a web interface have been announced.

The magic of Midjourney lies in advanced machine learning technologies. It operates based on LLMs and diffusion models. These models help Midjourney understand the text prompts' meaning, converting it into a numerical version or a vector. The diffusion model then guides the image generation process, starting with random noise and ending with high-quality artwork.

While some aspects of Midjourney's functionality remain unknown due to its closed-source nature, we know the tool employs a machine learning technique called diffusion, but it's unclear whether it's based on the open source Stable Diffusion model. It's important to note that Midjourney is a closed-source and proprietary tool.

The versions' evolution

Since its origin, the Midjourney team has been working on improving its algorithms so that new model versions are released every few months, most of them with groundbreaking features. At the time of writing, the V5.2 model is the latest and most advanced model, providing more detailed and sharper results. It was released in June 2023. However, rumor is that a V6 model is just around the corner.

Despite the evolution, you can still use the old models (*Figure 1.4*); you may want to use older versions simply to compare each image generated with the same prompt across different versions (to see how the model has improved over time) or to obtain specific results according to the characteristics of each model. You will learn more about this in *Chapter 3*.

Figure 1.4 – Images generated with the same prompt ("/imagine a beautiful blue flower in a vase by a window") input across different model versions

While Midjourney allows for the quick creation of digital images from text instructions, usage and ownership rights may pose a problem.

Legal challenges and ethical concerns around Midjourney

The emergence of AI art has sparked numerous debates over copyright ownership. The consensus leans toward granting copyright protection to outputs that are products of AI acting as a tool under human control. However, AI-created outputs with minimal human intervention are less likely to receive protection. The level of human intervention required for copyright protection remains a contentious issue and is subject to court deliberation.

Also, while you can use the images you create, they may be used by others for remixes. There's an ongoing debate about the legality of AI image generators such as Midjourney; despite these tools generating original images, they are trained on datasets that contain existing artwork. This has led to discussions about copyright infringement versus the fair use doctrine. Note that every image produced with Midjourney in any of the pay plans can be legally utilized by others, even for limited commercial purposes.

> **Copyright infringement with AI art**
>
> Copyright infringement occurs when someone uses copyrighted material without permission, violating the exclusive rights granted to the copyright holder. In the context of AI-generated art such as Midjourney, these concepts can become blurred, leading to legal ambiguity and debate.

In addition, there have been notable instances where Midjourney's use has stirred both applause and controversy. A Midjourney image titled *Théâtre D'opéra Spatial* (*Figure 1.5*) won the digital art competition at the 2022 Colorado State Fair, causing debate about the value of AI-generated art in such competitions. *The Economist* and the Italian newspaper *Corriere della Sera* used Midjourney-generated images for covers and comics, respectively, igniting discussions about AI's role in replacing human artists.

Figure 1.5 – Jason M. Allen/Midjourney: the Colorado State Fair 2022

In 2023, AI text-to-image generators such as Midjourney have gained even more popularity for creating realistic and occasionally controversial images. Images such as a fictional arrest of Donald Trump or a photo of Pope Francis in a white puffer coat went viral, demonstrating the power of these tools, as well as their potential misuse.

In conclusion, Midjourney is a promising game-changing tool that is redefining creativity and has the potential to revolutionize the creative industry. The implications, both positive and negative, are vast. On the one hand, it can democratize art, making it accessible to everyone, regardless of their artistic skill. On the other hand, it raises questions about the value and uniqueness of human-created art, copyright issues, and the potential for misuse. As with all technology, how it is used will ultimately determine its impact on society.

In this section, we explored Midjourney's underlying technology, evolution, and the associated legal and ethical debates. As we've seen, this revolutionary tool is already making waves in the creative industry. Now, let's turn our attention to a more concrete examination of its applications and influence. From boosting creativity and efficiency to reimagining the advertising landscape, we will explore the tangible benefits that Midjourney offers to artists, businesses, and even literature.

Exploring the benefits of Midjourney

As previously mentioned, Midjourney is revolutionizing various aspects of the creative industry. Its ability to convert text prompts into high-quality images significantly enhances productivity, with a wide range of applications in different sectors. Let's delve into the benefits of using Midjourney and examine some real-world examples.

First, Midjourney helps boost creativity and efficiency in the creative process. Midjourney, at the core of its functionality, presents nearly limitless possibilities for design and image generation. The tool, known for its efficiency, generates high-quality outputs in a fraction of the time compared to traditional methods (such as hand drawings, hand-made illustrations, and paintings, or using physical design tools or employing standard graphic design software such as Adobe Photoshop or Illustrator). Across the globe, numerous brands have integrated Midjourney into their design workflows, realizing its benefits and harnessing its capabilities to stretch the boundaries of creativity.

Midjourney also allows for quick prototyping and enhancement of creativity. As David Holz stated in an interview with *The Register*, "*The professionals are using it to supercharge their creative or communication process*" (https://www.theregister.com/2022/08/01/david_holz_midjourney/). Artists can employ Midjourney to swiftly develop prototypes of artistic concepts, such as mood boards, to present to clients before immersing themselves in their work. This method encourages the rapid generation of high-quality visuals for presentations, significantly accelerating the ideation and creation process.

I have personally utilized Midjourney as a brainstorming tool in my day-to-day work. It allows me to start with a vague concept and test various effects and iterations. Sometimes, the results unexpectedly form the basis for new ideas and projects. Using Midjourney as an ally and an extension of our creativity is, indeed, an excellent way to harness the potential of this tool.

Exploring the benefits of Midjourney   13

In terms of application, Midjourney has found significant traction in the advertising sector. Its capacity to rapidly generate unique content and brainstorm ideas presents a novel set of opportunities. These include the creation of customized ads for individuals, increased efficiency in e-commerce advertising and social media content, and innovative ways of creating special effects.

Baskin Robbins, the well-known ice cream brand, is a good example of a brand that has seen the potential of this tool. Teaming up with Tapan Aslot, a renowned AI artist who previously gained recognition for his *Food with a Golden Touch* series on Instagram and LinkedIn, the brand embarked on a groundbreaking journey. Together, they used Midjourney's GenAI capabilities to launch a new ice cream flavor campaign, based on an enchanting ice cream wonderland of AI-generated images. The unique, visually appealing creations sparked widespread interest on social media platforms, showcasing the brand's innovation in leveraging AI technology and the unparalleled potential of AI to revolutionize traditional marketing paradigms:

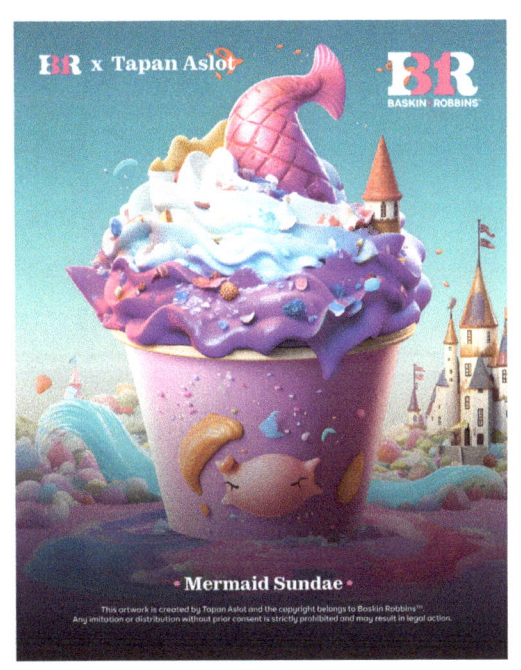

Figure 1.6 – Two examples of the Baskin Robbins flavor campaign by Tapan Aslot

Martini is another well-known brand that used this tool to remix its existing stock of promotional imagery to create a new AI campaign, *Unbottling Martini*, with a unique view of the Martini world.

Another unique application of Midjourney was seen in the AI-generated children's book, *Alice and Sparkle*:

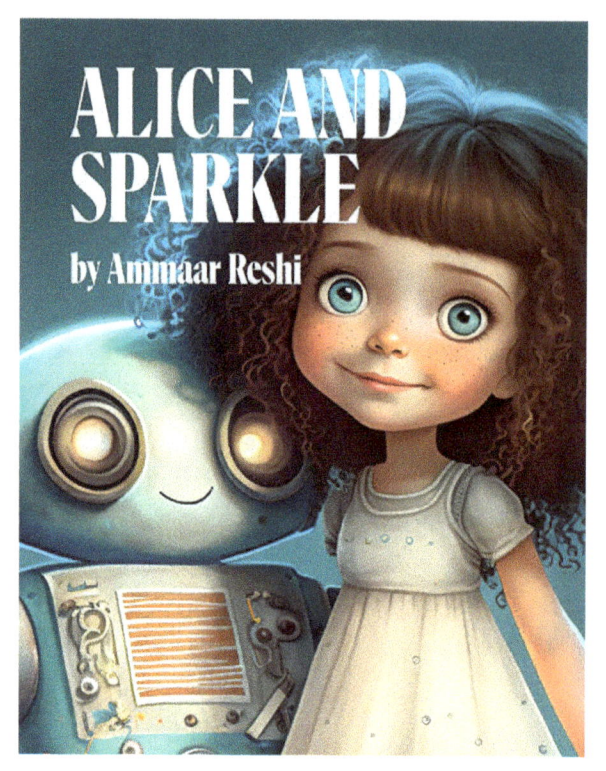

Figure 1.7 – Cover of the 2022 book Alice and Sparkle by Ammaar Reshi

The book's creator, Ammaar Reshi, spent hours refining Midjourney prompts to select the perfect illustrations. This innovative application showcased the potential of AI tools in literature

In conclusion, integrating Midjourney into your creative process can greatly enhance your output. With AI at its core, Midjourney provides a virtually limitless array of design possibilities. It's a resource-efficient tool that can generate high-quality designs in a fraction of the time that traditional methods would require. Many brands globally have already incorporated Midjourney into their design workflows, realizing its benefits and harnessing the tool to push creative boundaries.

Summary

In this chapter, we embarked on a journey to examine the transformative influence of AI in the creative industry and debated its role, focusing on the AI-powered tool, Midjourney. We started with a broader understanding of AI's implications and then proceeded to explore the world of generative art, discussing AI technologies that drive it, such as GANs, VAEs, and diffusion models. We touched upon future challenges and opportunities of AI in the realm of art.

We took an in-depth look at Midjourney, discussing its origins, underlying technology, and evolution. We highlighted its myriad benefits and applications across various sectors and presented real-world examples of its use. We also explored the controversy surrounding AI-generated art and its impact on copyright and ethical debates.

The insights gleaned from this chapter offer a holistic understanding of the fusion of AI and generative art, with a concentrated emphasis on Midjourney. As we venture into the next chapter, *Embarking on a Journey – From Discord to Midjourney*, we will equip you with practical knowledge to set up and use Midjourney via Discord. This essential knowledge will help you embark on your own journey of AI-driven art creation with Midjourney.

2
Embarking on a Journey – From Discord to Midjourney

The fusion of technology and creativity has paved the way for extraordinary tools such as **Midjourney**, and now it's time to explore how you can use **Discord**, a free voice and text social platform, to enhance your AI art creation journey. This chapter serves as your gateway into the world of Discord in connection with Midjourney, guiding you step by step through the process of creating a Discord and Midjourney account to having your own Discord server.

Whether you are a complete beginner or seeking ways to optimize your Midjourney setup, this chapter provides essential instructions on how to join Midjourney, install and use Discord, invite the Midjourney Bot to your server, and much more. With practical insights, you'll be able to use Midjourney through Discord on your own server.

In this chapter, we're going to cover the following topics:

- **Diving into Discord**: An introductory guide to Discord, a versatile platform that serves as the gateway to accessing Midjourney. In this section, we will cover how to sign up for Discord, understand its features, and how it facilitates the creative process with Midjourney.

- **Logging in to Midjourney and connecting to Discord**: A step-by-step guide on how to join Midjourney through Discord, including setting up your Midjourney account and linking it with Discord. This segment aims to smooth your entry into the world of AI-generated art, ensuring an efficient setup process.

- **Inviting the Midjourney Bot to your server**: You'll get **a** detailed walkthrough on how to enhance your creative workflow by inviting the Midjourney Bot into your own Discord server. This enables a personalized and organized environment for art creation, making the process more efficient and tailored to your needs.

By the end of this chapter, you will be skilled at using Discord with Midjourney, opening up a world of possibilities for smoother and more personalized art creation. You'll understand how to maximize the features of Midjourney, ensuring a productive and enjoyable creative process. Ready to embark on this exciting journey? Let's dive in!

How to join the adventure

Creating an AI-powered masterpiece, building your portfolio of artistic visuals, and unlocking your potential with Midjourney's AI art generative tool is just the beginning of what awaits you. Let's embark on this journey together!

First, we need to access Midjourney. Interestingly, the only way to access Midjourney is through Discord. So, what exactly is Discord? Discord is a communication platform that was originally designed for gamers, but it is now widely embraced by various communities. It offers text, voice, and video chat capabilities, allowing real-time connection across servers centered around specific topics or interests. Bots enhance Discord's functionality, providing automation and integration with platforms such as Spotify or YouTube. Moreover, custom bots can be created using Discord's API. The platform's versatility supports casual social interactions and organized group activities, making it a go-to place for community engagement.

So, as we know, Midjourney's AI art generator tool is an innovative tool designed to foster creativity. As previously mentioned, through the combination of AI and machine learning, you can generate stunning visuals without spending countless hours on manual design. As mentioned, the only way to access Midjourney is through Discord, which offers a user-friendly interface, perfect for your artistic explorations. Discord can be used for free in two ways: by accessing it directly in your browser or by downloading the app for use on desktop, mobile, and tablet.

In brief, here's what we're going to be doing in this chapter:

- Signing up for Discord
- Signing up for Midjourney and picking our Midjourney plan
- Jumping to Discord and accepting the invitation to the Midjourney Discord channel
- Start generating and creating AI art

Diving into Discord

Let's take a look at how you can get started with Discord through your browser (keep in mind that you can also download the Discord app).

First, let's sign up for Discord:

1. Go to the Discord home page (www.discord.com) and select **Login**:

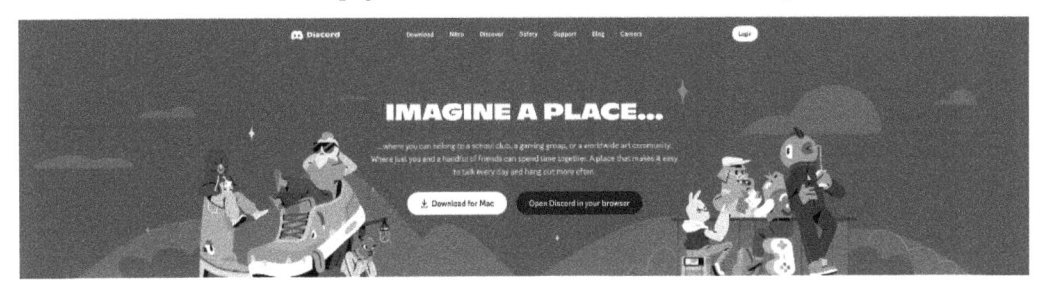

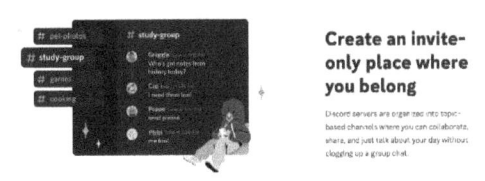

Figure 2.1 – Discord home page

2. Create your account on the registration page:

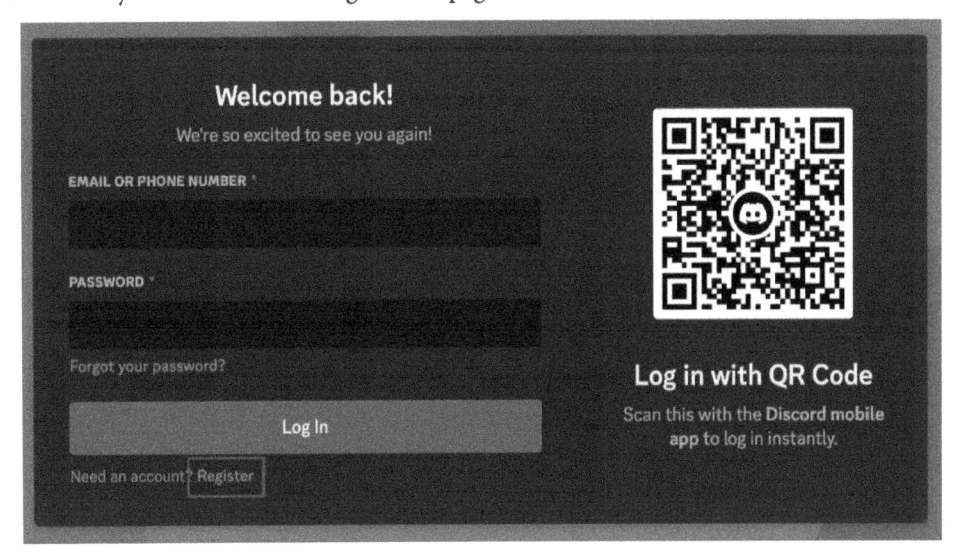

Figure 2.2 – Discord's create an account form

3. Select **Register**, and then complete the form by entering your email address, username, password, and your date of birth. Accept the terms of use and click **Continue**:

Figure 2.3 – Discord's create an account form

4. After this, a verification email or phone message will be sent to you. This is a safety measure that helps prove you are not a robot and helps keep spammers out. Just click on **Verify Email** in the email message you received:

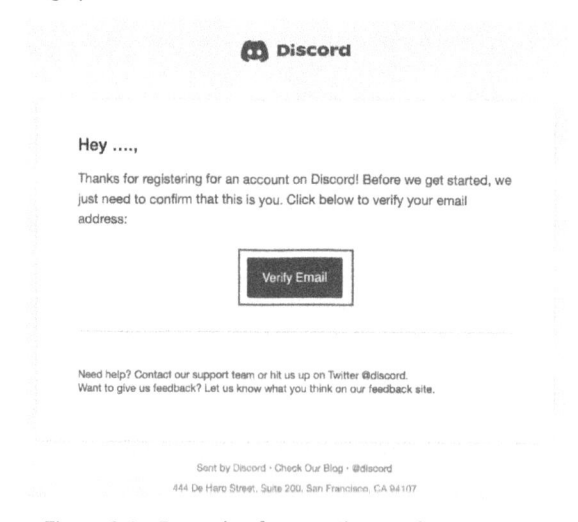

Figure 2.4 – Example of an email to verify your details

Now that your registration is completed, you can log in via the Discord website in your browser or download the app on the Discord homepage (*Figure 2.1*) for use on mobile, tablet, desktop, or laptop. The app is free of charge and available for Windows, Linux, Mac, Android, and iOS.

Congratulations! You're now set up on Discord and are ready to explore the world of AI art creation.

Next, we'll dive into how to connect your Discord account to Midjourney, unlocking a universe of creative possibilities. Keep reading to discover how to make the most of these powerful tools.

Logging in to Midjourney and connecting to Discord

With your Discord account ready, the next step is to plunge into the world of AI-generated art with Midjourney. Here's a comprehensive guide to getting started:

1. Sign up for Midjourney.

    To do so, visit the Midjourney home page at www.midjourney.com and click on **Join the Beta**:

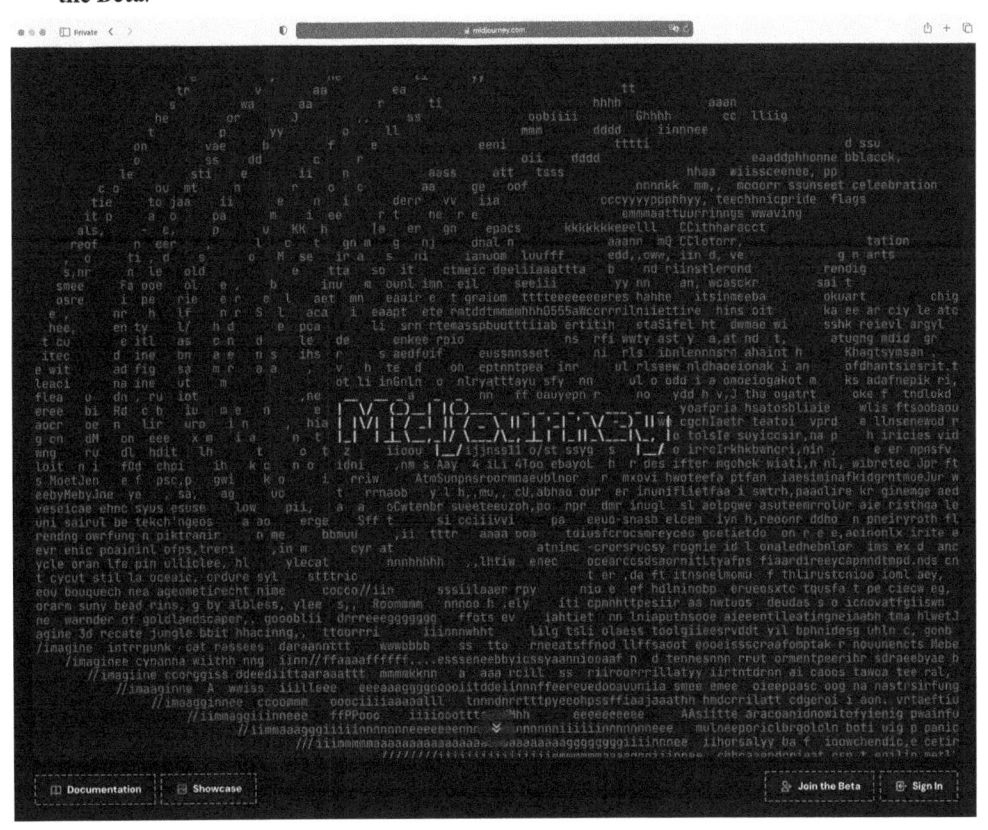

Figure 2.5 – Midjourney home page

2. This will open a page with an invite to join the Midjourney Discord channel. Click **Accept Invite**:

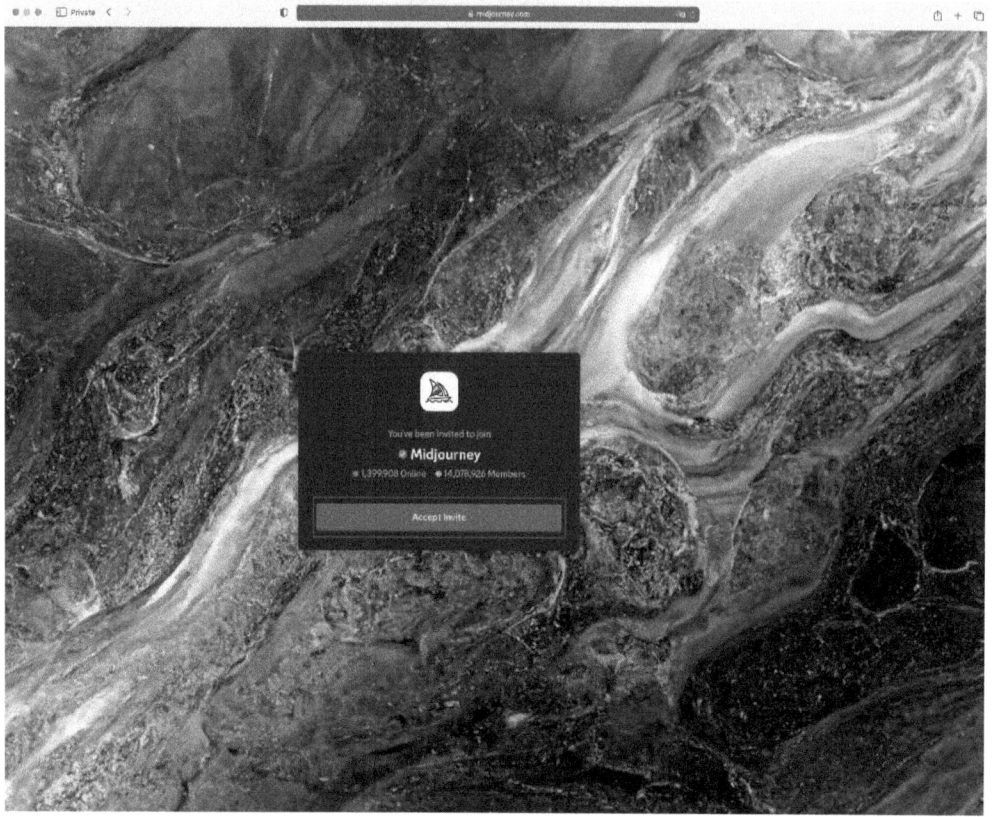

Figure 2.6 – Midjourney invite to join

Enter a username of your choice and select **Continue**.

3. After you have logged in with your verified account on Discord, go back to Midjourney's home page and click the **Sign in** button (*Figure 2.5*); you will be redirected to a page where you just have to select **Authorize** to authorize the access:

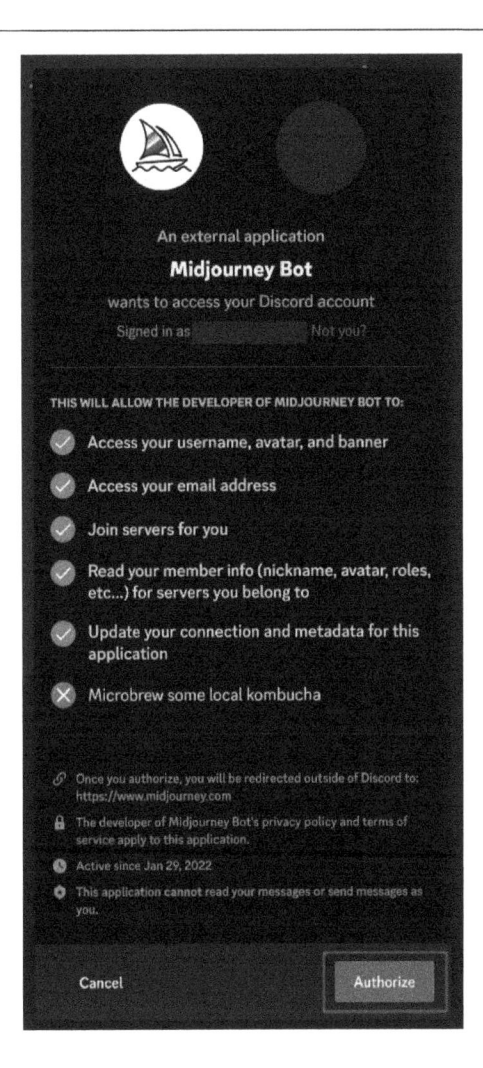

Figure 2.7 – After clicking Sign in, you will need to authorize the
Midjourney Bot to access your Discord account

4. You will now be able to access your Midjourney profile. Here, you can manage your subscription through the **Manage Subscription** button in the left side menu, and choose the plan that works best for you.

Currently, Midjourney does not have a free plan. In my opinion, the *Standard* plan is ideal, either to get started or to use on a more daily basis. There's quick and easy management of the hours that can be used (the pricing packages are stipulated by GPU hours, for example, each *fast hour* can generate around 60 images). With the *Basic* plan, you are limited to more or less 200 generations per month, and with the *Standard* plan, the management can be done between the use of 15 hours in *fast hours* mode and unlimited generations in *Relax* mode, which takes a little longer to generate images. In any case, we'll talk about these details and how to configure them in the next chapter.

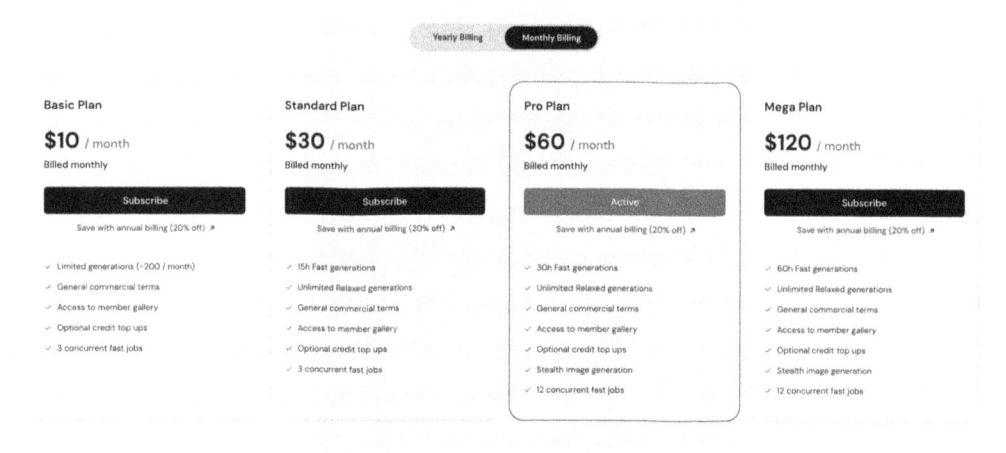

Figure 2.8 – Manage subscription Midjourney page

Manage subscription tip

Another way to go to your subscription page when in Discord is to go to one of the `#newbies` channels, and type **/subscribe**. This is a slash command and it's how you interact with Discord bots such as Midjourney. As you enter this command, a link will appear that will direct you to the Midjourney plans.

How to join the adventure 25

On the Midjourney home page (midjourney.com) (*Figure 2.5*), I advise you to visit the **Documentation** link, where you can find all the detailed information about Midjourney:

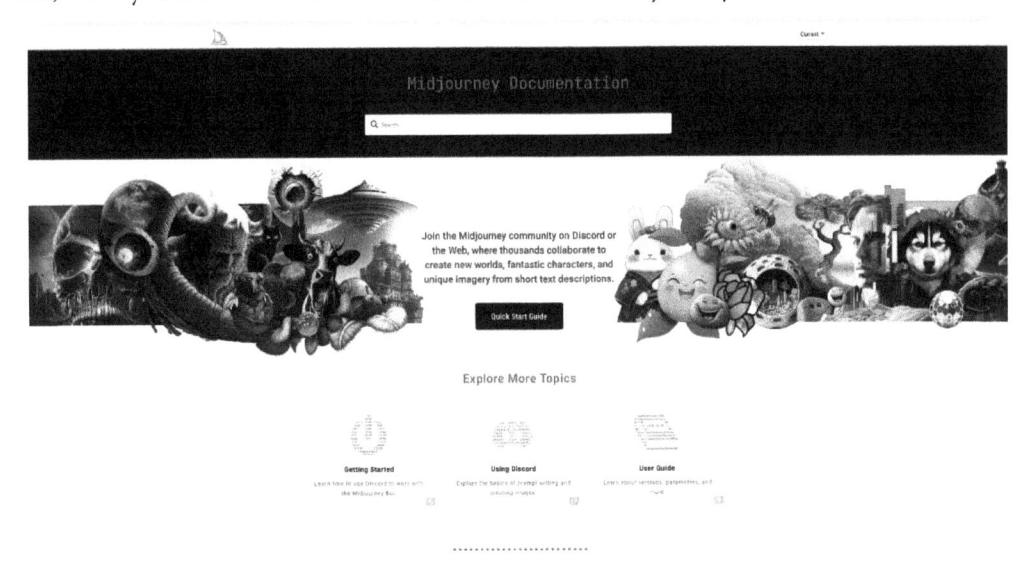

Figure 2.9 – The Midjourney Documentation page

I would also encourage you to click the **Showcase** button on the home page to view a small showcase of user-generated art:

Figure 2.10 – Midjourney Community Showcase for inspiration

Now that you have your Discord account, chosen your Midjourney subscription plan, and invited the Midjourney Bot into Discord, it's time to start interacting with the Midjourney Bot. Once you're in the Midjourney server on Discord, locate and select any channel labeled #general or #newbie:

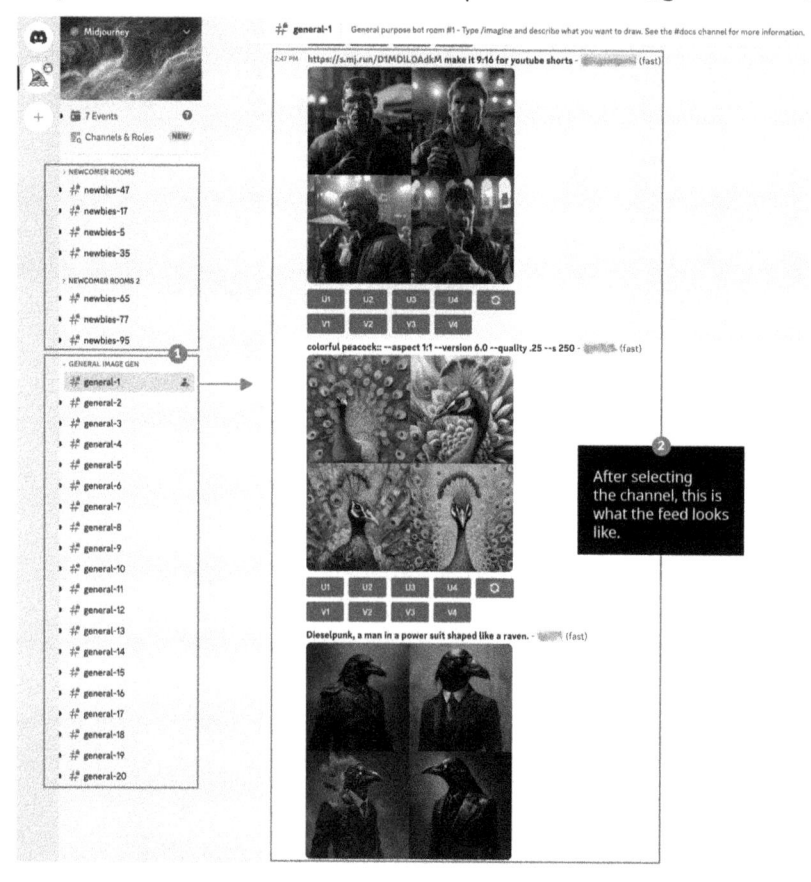

Figure 2.11 – Midjourney channels to start with and the appearance of the channel feed

Once in a #newbie or #general channel, the only step that follows is to start prompting. You can do that just by typing the command **/imagine** followed by a text description of your choice, for example, **/imagine** A photorealistic view of a sunny day in the mountains. Send your message and the bot will start generating your image. I recommend spending a little time testing some different prompt concepts and looking for what other users are also prompting; it's a wonderful way to learn and come up with ideas.

Figure 2.12 – Interacting with the Midjourney Bot to generate an image

Over time, you will start to see that it can start to be very chaotic to be using these public channels. Normally, these channels are very busy, and it can be hard to find the images generated by your prompt. So, next, we will learn how to invite the Midjourney Bot to your server, enabling a more organized and personal space for your AI-generated art, and start prompting in a quieter place.

Want quieter? Inviting the Midjourney Bot to your server

Now that you've mastered working with the Midjourney Bot on Discord, it's time to elevate your experience and streamline your creation space. If you've ventured into the `#newbies` and `#general` channels, you may have found yourself lost in the torrent of results from thousands of fellow users making constant prompts. That's where inviting the Midjourney Bot to your server comes into play (you can generate images with the Midjourney Bot on any Discord server that has invited the Midjourney Bot). Not only is this the perfect solution to declutter your prompt results, but it's also an incredibly straightforward process, as you'll discover soon.

Let´s start by following these few steps:

1. Create your own server.
2. Invite the Midjourney Bot to your server.
3. Start prompting. Now, you are ready to begin. We only introduce the basics here; we'll explore this topic in detail in the next chapter, where you'll start generating your first images. For comprehensive insights on prompting, see *Chapter 3*.

Creating a server

The best way to create your own server is by clicking the + icon on the left column of the Discord panel:

Figure 2.13 – Click the + button on the left-hand column

After this, a new window will appear with four options – **Create My Own**, **Gaming**, **School Club**, and **Study Group**. We will choose the first option, **Create My Own**. The next window will ask you more details about your server; you can hit **skip this question**, following which you will finally be asked to **Customize your server** with a name and a profile picture (optional). Then, click **Create**. That's it – you have your first server!

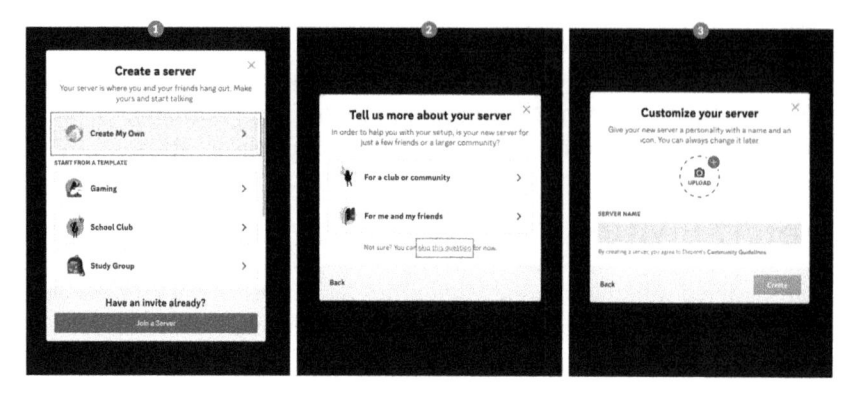

Figure 2.14 – Steps to create your server

Inviting the Midjourney Bot to your server

After setting up your server, go to the Midjourney Discord server and select the Midjourney Bot from the member list; in case you don't see the member list, click on the member list icon (*step 1* in *Figure 2.15*) on the top right of the menu. After selecting the Midjourney Bot (*step 2* in *Figure 2.15*), a popup will appear; click on **Add App** (*step 3* in *Figure 2.15*):

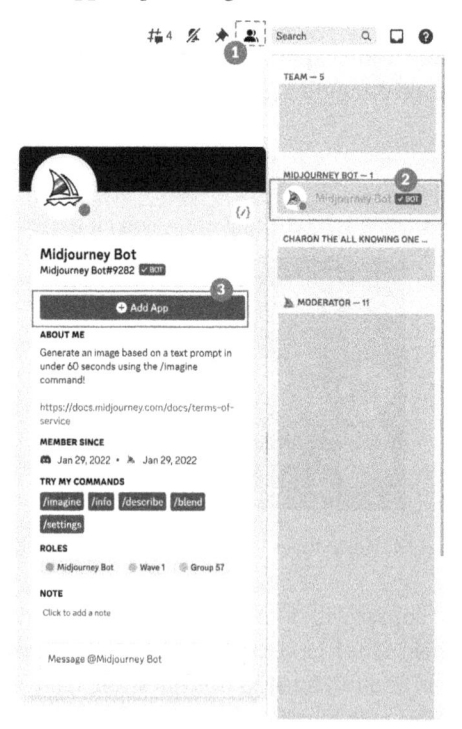

Figure 2.15 – Steps to add the Midjourney Bot to your server

A new popup will appear, and you only need to choose the server that you want to add the Midjourney Bot to, select **Continue**, and then select **Authorize**.

Now, it's time for the final, yet vital stage of organization. Within **TEXT CHANNELS**, consider creating distinct channels by clicking on the + sign, each dedicated to specific themes you plan to explore:

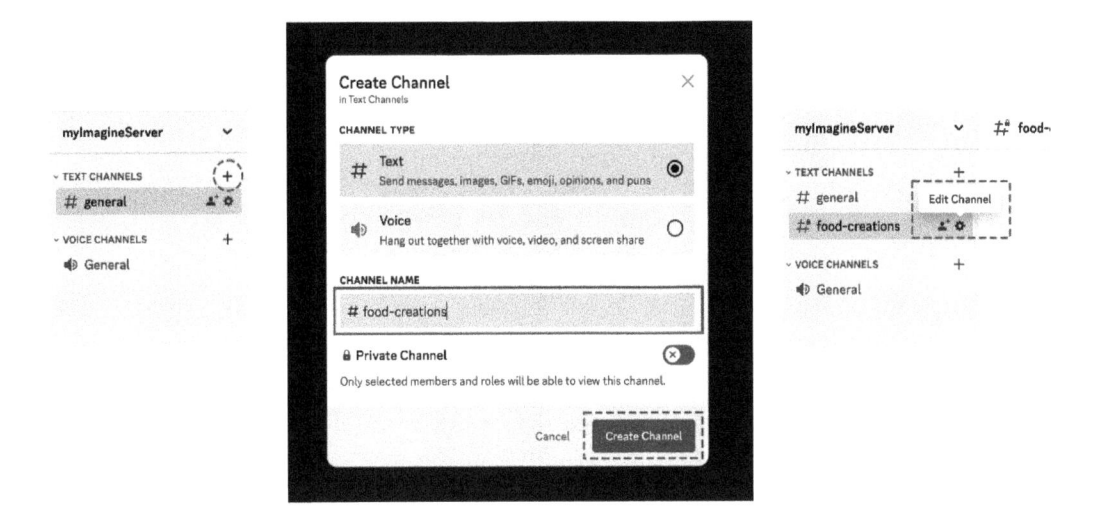

Figure 2.16 – See how you create new Text Channels by clicking the + button and how to edit them

Whether you choose to categorize them by subjects such as food, landscapes, animals, or even versions of the Midjourney model you want to experiment with (more on this later), this tailored structure will make locating specific images a breeze in the future. Your organizational design is personal to you, offering an efficient way to enhance your creativity and keep your ideas neatly arranged.

Once you've completed the preceding steps, the Midjourney Bot will be up and running on your server.

Start prompting

To test the Midjourney Bot, click on your server, choose a channel from the **TEXT CHANNEL** list, such as #general, and enter the **/imagine** command followed by your text prompt:

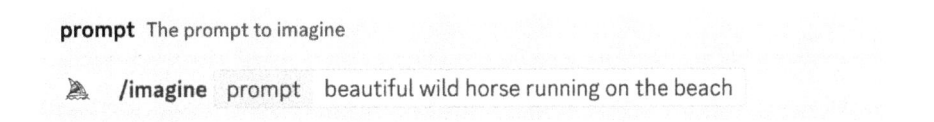

Figure 2.17 – Typing /imagine in the message box

I will use **/imagine prompt** `beautiful wild horse running on the beach`:

Figure 2.18 – Example of the result of my prompt – if you have used the same prompt as me, your images will be similar to this but never equal to them

In conclusion, you've successfully taken control of your creative environment by inviting the Midjourney Bot to your own server. Now, free from the chaos of crowded channels, you have a clean and organized digital canvas waiting for your artistic explorations. These steps have prepared you for a unique journey. Your own server, with its neat and dedicated channels, is ready for you to explore, create, and innovate. The only limit is your creativity.

Summary

In this chapter, we set out on a journey to explore the convergence of Discord and Midjourney, opening doors to a new era of AI-generated art creation. We began with an introduction to Discord, detailing its functionalities and highlighting its essential role as the gateway to Midjourney. Next, we provided guidance on how to join Midjourney, set up an account, and select an appropriate plan.

We then guided you through the process of inviting the Midjourney Bot into Discord, offering systematic instructions to navigate the setup. Additionally, we explained how to create a personalized server for a more focused and organized art creation space, along with the methods to invite Midjourney Bot to your server.

Through hands-on lessons and practical insights, this chapter has readied readers for their unique creative paths, arming them with the knowledge and tools required to make the most of the powerful union of Midjourney and Discord.

In the next chapter, we will start exploring the world of prompts, mastering commands, enhancing prompts, and fully unlocking the extraordinary capabilities that Midjourney offers across its various model versions, taking creativity to unprecedented levels.

Part 2: Unlocking the Power of Midjourney – A Deep Dive into Features and Functionalities

In this part, we dive into the expansive features and functionalities of Midjourney. This section is designed to unlock the full potential of Midjourney, guiding readers through the various versions of the tool and providing a comprehensive understanding of how to make use of its capabilities to create visually stunning AI art. From mastering different versions to exploring upscalers and parameters, readers will gain the insights needed to navigate Midjourney's rich feature set effectively.

This part has the following chapters:

- *Chapter 3, Mastering Midjourney Versions – Quick Start Guide*
- *Chapter 4, Understanding and Learning Parameters*

3
Mastering Midjourney Versions – Quick Start Guide

Reflecting on our prior exploration, we've discovered the transformative influence of AI on creativity and grasped the nuances of setting up Midjourney. Now, as we progress, we'll dive deeper, much like a musician understanding their instruments. Imagine having an array of musical tools at your disposal. Knowing the distinct sounds and characteristics of each can profoundly impact the final symphony. In a similar vein, understanding the evolving iterations of Midjourney allows artists and tech enthusiasts to produce content that resonates best with their creative vision.

Each version of Midjourney, from its inception to its recent advancements, represents a milestone in technology and possibilities. Knowledge of the Midjourney versions isn't just about utilizing a tool; it's a deep dive into its journey, understanding how each evolution can impact your creative pursuits.

In this chapter, we're going to cover the following topics:

- **Imagine and prompt**: The foundational step to any Midjourney project. This section breaks down the art and science behind crafting effective prompts that communicate your ideas to the AI, transforming words into compelling visual outputs.
- **My first prompt – now what?**: What comes after taking that initial step? Here, we explore the immediate steps to take after your first prompt, including understanding the output and how to refine it.
- **Beyond /imagine – command list**: Expanding your horizons with the myriad commands at your fingertips. Here, we explore the comprehensive list of commands available in Midjourney, enhancing your ability to navigate and use the platform to its full potential.
- **Model versions 1 and 2**: Revisiting the origins and understanding the initial breakthroughs. An examination of the early versions of Midjourney, that laid the groundwork for what would become a revolution in AI art.

- **Model versions 3 and 4**: The advancements that set newer benchmarks. This section discusses the improvements and new features introduced in these versions, offering insights into how they enhance the creative process.
- **Model versions 5 to 5.2**: Understanding the pinnacle of Midjourney's innovations, we explore the latest and most advanced versions of Midjourney, with the most detailed and realistic outputs we have seen so far.
- **Niji versions – the anime world**: A side journey into a model that brings its own set of wonders. Focused on the Niji versions, this part uncovers how Midjourney approaches the creation of anime-style art, showcasing its unique features.

By the end of this chapter, you'll have a comprehensive grasp of each of these versions, enabling you to select and deploy the one that best aligns with your project's goals. Why explore the intricacies of each version? The nuances, the capabilities, the differences –each has its unique value, ensuring your projects aren't just good but exceptional.

Technical requirements

To make the most of this chapter and follow along with the hands-on activities, you'll need to ensure you've set up everything we covered in *Chapter 2*. Here's a quick recap:

- **Midjourney account**: You'll need an active account on Midjourney. If you haven't registered yet, refer back to *Chapter 2* for a step-by-step guide.
- **Discord installation**: Ensure you have Discord installed on your device since we'll be working with it alongside Midjourney. Again, the installation guide is available in *Chapter 2*.
- **Payment plans**: At the time of writing, Midjourney has no free or trial plan. A Midjourney paid plan is necessary. Please ensure you've selected the most suitable plan for your needs.
- **Server creation**: Having your own server can significantly enhance and expedite your experience with Midjourney. If you're yet to set up a server in Discord, *Chapter 2* covers the process in detail.

Imagine and prompt

As you explore AI art in more detail, it will become clear that creating a precise and effective prompt is a skill that is central to the process of generating what you intend. The connection between your imagination and the AI's visual representation transforms simple words into captivating images. As we proceed through this section, we'll provide you with the following:

- An understanding of the art of AI prompt construction
- Proficiency in the structure and elements of influential prompts

While Midjourney can effortlessly churn out visuals based on prompts, the quality and relevance of the generated image heavily rely on how effectively the prompt is crafted. It's not just about telling the AI what to do; it's about communicating a vision.

Midjourney **prompts** are no ordinary strings of text. They are structured commands that guide the AI, much like how a compass guides a ship, ensuring the resulting visuals align with your envisioned picture. There are several ways to structure a prompt, but over time, I've noticed that the most effective is the simplest one; the basic structure might have the following three aspects:

- *Subject* (this is where you specify the main subject of your image, and it can be a building or a city, an animal, a person, and so on)
- *Action* (this is the context of the action, whether it's a building on a busy street, a city with Christmas lights, a dog running on the beach, or a person in a classroom, and so on)
- *Details* (specify or detail more information you want to include, such as backgrounds, environments, moods, and so on)

Although Midjourney is one of the best tools available and can deliver extraordinary results with just the use of one-word prompts, emojis, or even images (more on this later), you'll notice that if you want to have more control over the final result, you should detail your prompt with essential attributes. To do this, you can also add the following to the basic prompt:

- Artistic styles, colors, and potential artist references
- Supplementary settings such as lighting, camera shot, and aspect ratios

Using this information regarding the structure of prompts, let's create our first image by following these simple steps:

1. Type **/imagine** in the message box:

Figure 3.1 – Typing /imagine in the message box

2. Start typing the basic prompt with your structured text prompt. In this case, let's try "*[subject]* `Golden retriever` *[action]* `playing fetch` *[details]* `on a sunlit beach with gentle waves`" and press *enter*:

prompt The prompt to imagine

/imagine prompt Golden retriever playing fetch on a sunlit beach with gentle waves

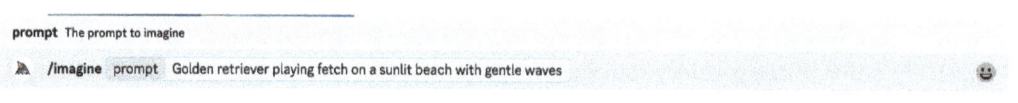

Figure 3.2 – Basic text prompt

3. Next, let's retype the same prompt, but this time, we'll be a bit more descriptive:

Again, type **/imagine** followed by "*[attribute]* `Polaroid Photo,` *[subject]* `Golden retriever` *[action]* `playing fetch` *[details]* `on a sunlit beach with gentle waves,` *[attribute]* `Silhouette lighting`":

prompt The prompt to imagine

/imagine prompt Polaroid Photo, Golden retriever playing fetch on a sunlit beach with gentle waves, Silhouette lighting

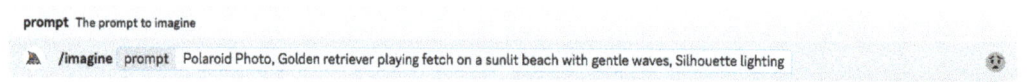

Figure 3.3 – Basic text prompt + more attributes

After a few seconds, the images will appear. The difference between the results is striking:

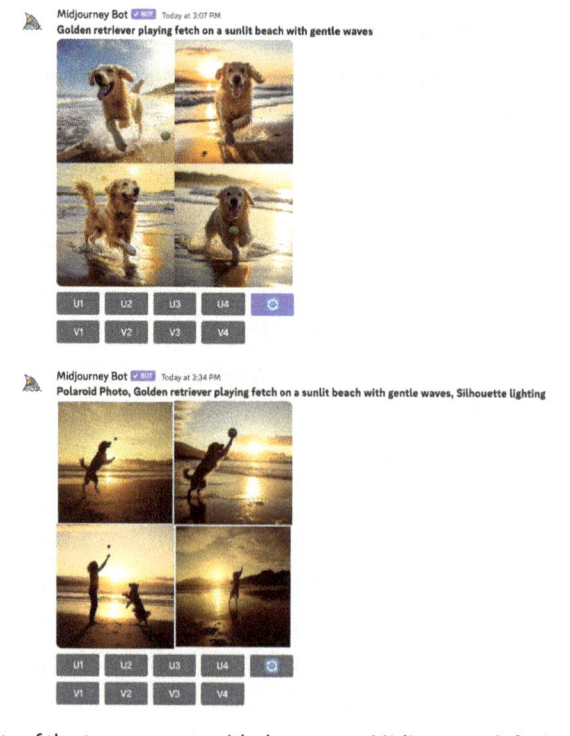

Figure 3.4 – Results of the two prompts with the current Midjourney default model version 5.2

For each prompt, Midjourney generates four images with nine buttons below those images; let's see what they are for and how we can use them.

My first prompt – now what?

Every time a prompt is sent to the Midjourney Bot, four images are generated (note that even using the same prompt, Midjourney will never generate the same image more than once). By default, the images that are created are generated in the latest Midjourney model, which is 5.2 at the time of writing.

Below the four-image grid are two rows of buttons:

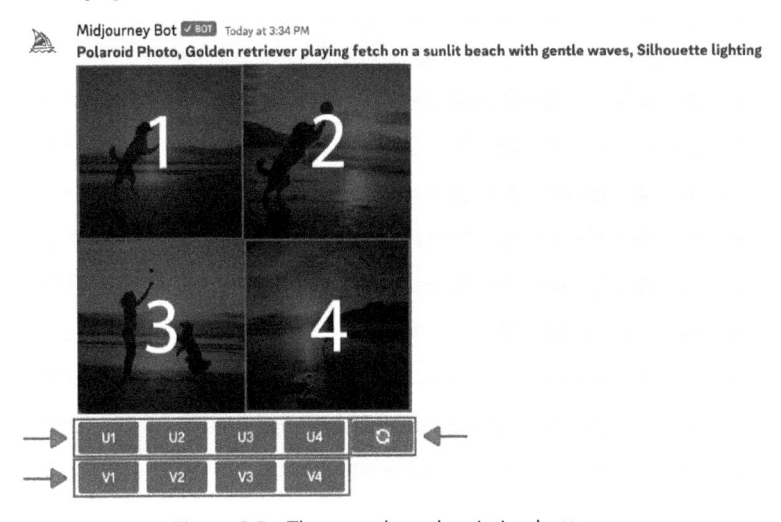

Figure 3.5 – The upscale and variation buttons

The numbers following the **U** (upscale) and **V** (variations) letters match the images: **1** and **2** for the top images and **3** and **4** for those below. In earlier Midjourney models, these images were produced in low resolution, so the **U1** to **U4** buttons were used to upscale our chosen image to a higher resolution. Since V5, images generated in the grid are now at their highest resolution, 1024 x 1024 pixels, approximately 36 cm x 36 cm at 72 DPI (more) on this in *Chapter 9*). The **U** buttons serve to isolate the chosen image(s) to get access to more functions and editing options: choose the image you prefer by clicking its corresponding button.

The **V1**, **V2**, **V3**, and **V4** buttons generate variations of your selected image, keeping the same style and composition intact. Clicking on any of these will present a new grid of four images. So, for example, if we click on **V1**, Midjourney will generate a new grid of four new images while retaining the same style and composition as *image 1* – that is, the one in the top-left corner.

The last button on the right is for re-rolling or re-running a job. It's frequently used to generate new results with the same original prompt.

You can re-roll and vary images endlessly, but after some attempts, the results might stray from your initial vision. If the desired image remains elusive after multiple tries, consider refining your prompt or altering the word order for a fresh grid.

Let's select one of the images to see what other options appear and what they are used for. In my case, I'm going to select *image 3*, by clicking on **U3**, from the prompt where I used attributes:

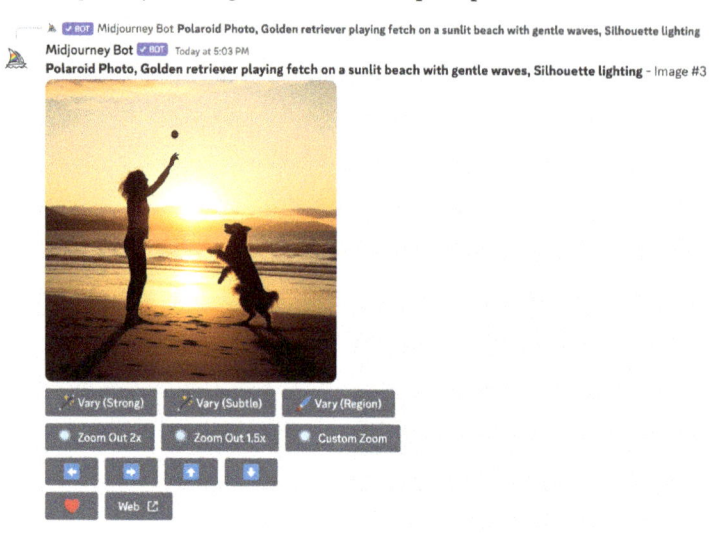

Figure 3.6 – The variation, zoom, and pan features

With the latest model version of Midjourney, you have extra buttons:

- **Vary (Strong)**: This is similar to the **V** button, but in this case, it creates four strong variations of the image. This can result in changes to the composition, elements, colors, and more.

- **Vary (Subtle)**: This is similar to the **V** button, but in this case, it creates four subtle variations of the image, preserving its original composition. This can be perfect for refining the result:

Figure 3.7 – Vary (Subtle) versus Vary (Strong)

- **Vary (Region)**: This feature allows you to edit regions of the image. It works best on large regions (between 20% to 50%) of the generated image, and it can be used with **Remix mode** to add more details or modify the prompt for that region. To use this feature, follow these steps (*Figure 3.8*):

    I. Press the **Vary (Region)** button; a popup will appear.

    II. Choose the rectangular or freehand selection tools at the bottom left and select the area of your image that you want to modify. If you enter a wrong selection, you can press the **Undo** button (the circular arrow in the top-left corner) to clear the selection.

    III. After selecting the area, in the text area of the message box next to the rectangular and freehand selection tools, insert what you want to add (this is only possible if you have *Remix* mode active; otherwise, the text area won't appear and after selecting the area, you'll jump straight to the submit step, at which point Midjourney will generate in *Free* mode only based on your image details). In my case, I'm going to write `skirt`.

    IV. Press **Submit** and close the pop-up editor.

> **Remix mode**
>
> Use *Remix* mode to alter prompts, adjust parameters, switch model versions, and modify aspect ratios amid variations. This mode considers the foundational composition of your initial image and incorporates it into the new job. Whether you're aiming to modify the ambiance or lighting, develop a subject, or craft new compositions, *Remix* mode is your ally. Activate it by typing **/prefer remix** in the message box and hitting *enter*. This command toggles *Remix* mode on and off. More details on this in *Chapter 7*.

Figure 3.8 – Vary (Region) with Remix mode active to change the image, adding a skirt to the lady

- **Zoom Out 2x** and **Zoom Out 1.5x**: These options zoom your image out and automatically generate details and content based on the original prompt.
- **Custom Zoom**: This gives you a pop-up text box that allows you to zoom out your image, add new details to the prompt, and change the aspect ratio. For example, you can set `--zoom 1` and then add `--ar 16:9` (the *aspect ratio* is a parameter in Midjourney that is initiated using double hyphens. This is the beginning step in declaring a parameter for your prompt. More about this in the next chapter):

Figure 3.9 – Example of Custom Zoom applied over the Vary (Region)-generated image

- **Pan Arrows**: This option will extend your image in the direction of the arrow (left, right, up, or down) without changing your original image. As a result, it's possible to expand your image to significantly larger dimensions and even enhance its resolution. For a detailed exploration of how image resolutions and sizes are manipulated within Midjourney, please see *Chapter 9*. At the time of writing, there is no limit to the number of times an image can be panned. However, after a certain number of pannings, typically around 12, the size of the resulting image may surpass what Discord can display due to its large dimensions. In such cases, you will receive a link to view your enlarged image in a web browser, thus overcoming Discord's display constraints. Typically, an image with a 1:1 aspect ratio, after undergoing 12 horizontal pannings, could reach dimensions of 7680 x 1024 pixels; however, it's important to notice that after panning an image, you can only pan the image again using the same direction (horizontal or vertical). As panning supports *Remix* mode, you can also boost your new extended images by changing your prompt to add more details:

Figure 3.10 – Example of the use of Pan Right with Remix mode
active to add more context to the new generation

- **Make Square**: This adapts the image to a new grid of four images but in square format (the base format is 1024 x 1024 pixels):

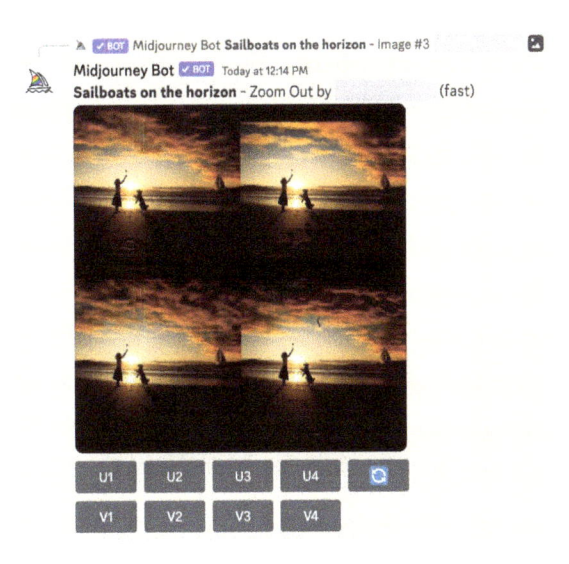

Figure 3.11 – Example of the Make Square option

- **Heart**: The heart is a default emoji; you can press it to indicate you liked a certain image. It can be useful when you're organizing your images on your Midjourney gallery page. To do this, simply go to the Midjourney site (`www.midjourney.com`), sign in, and click on the **Archive** button on the left side, navigate to **Filters** on the right side of of the screen, search for **Rating,** and choose **Liked**. Now, only the images that received a heart reaction will be shown:

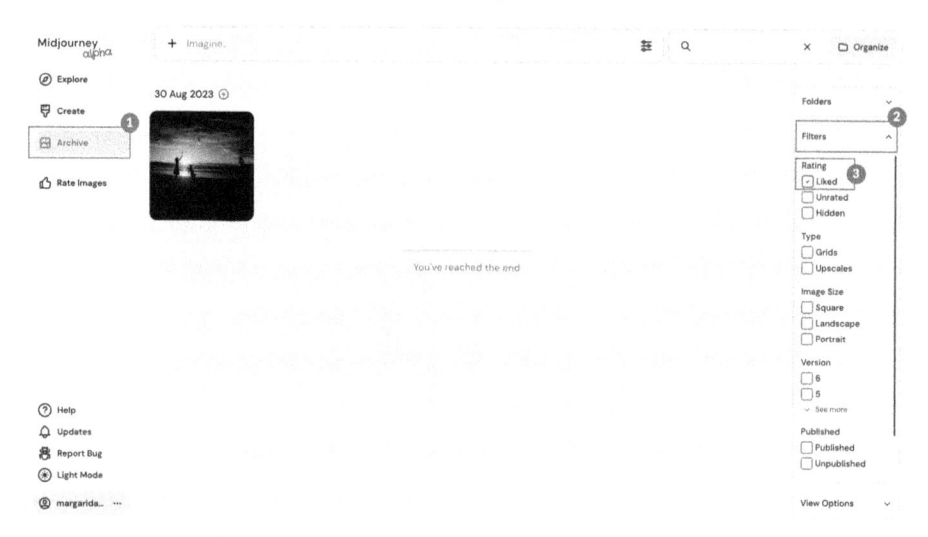

Figure 3.12 – Liked option on your Midjourney page

- **Web**: This button will open a pop-up box informing you that you are leaving Discord and will open your image in your Midjourney gallery:

Figure 3.13 – This is what the final image looks like and appears in your personal Midjourney area

Having explored the fundamental functionalities of Midjourney, we now possess a clear understanding of its image-generation process. From the image grid layout to the diverse **Vary** features and the detailed workings of *Remix* mode, we've covered essential elements such as zooming and panning.

With this foundational knowledge in hand, we can move forward to our next section, *Beyond /imagine – command list*. This section will present a comprehensive list of every command available within the platform, further enhancing your familiarity with Midjourney's features. As we progress, be ready to deepen your skills and knowledge of the system.

Beyond /imagine – command list

Now that we've had our first interactions with Midjourney and have begun to understand the mechanics of how it works, let's explore the existing commands that will not only enhance our workflow but will also allow us to personalize our preferences.

You can interact with the Midjourney Bot on Discord through these commands, which serve to produce images (as you saw earlier with the **/imagine** command), modify settings, monitor user information, and more. They're applicable in any bot channel, on private Discord servers where the Midjourney Bot is permitted, or when directly messaging the Midjourney Bot. To trigger any type of command, simply type /, followed by the command in question, in the message box and hit *enter*. It's important to note that Midjourney interprets these commands with a high degree of precision and does not tolerate any mistakes, such as grammatical or spelling errors. This strictness is in contrast to the more lenient handling of queries by typical search engines.

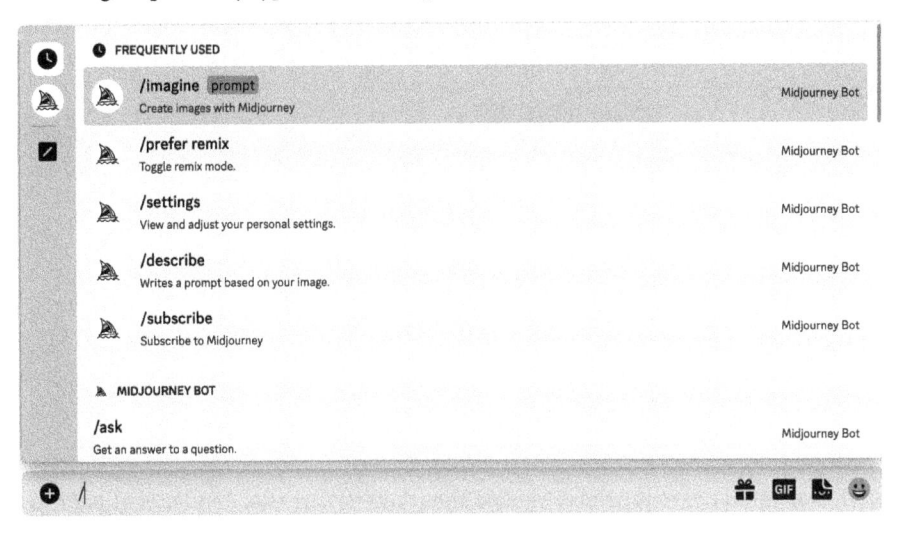

Figure 3.14 – Example of some existing commands that appear as soon as you type /

Direct messages

As a Midjourney subscriber, you can have a one-on-one chat with the Midjourney Bot via Discord's **Direct Messages**. Remember, images created in this way are still under content and moderation guidelines and will appear in your Midjourney website gallery. To start a conversation with the bot, do the following:

1. Find and click on the Midjourney Bot's name, either in the **Member List** area or wherever it's mentioned.

2. Initiate the conversation by sending any message to the bot.

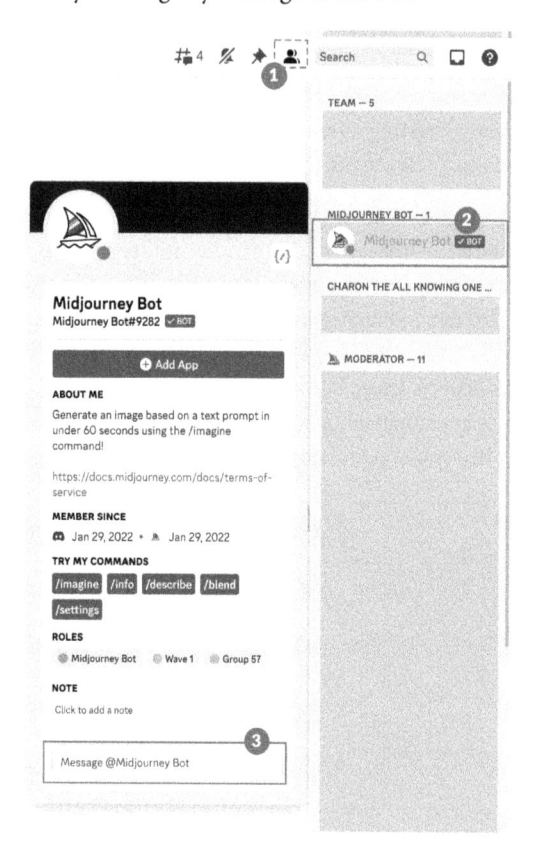

Figure 3.15 – Where to find the bot and how to message it

Here is a complete list of the commands available and what they are used for. Throughout this book, we will explore the more complex commands and how to fit them together to get the best results for your specific needs:

- **/ask**: Retrieve an answer to a specified question.

- **/blend**: Combine up to five images to create a unique and original one. More on this in *Chapter 5*.
- **/daily_theme**: Toggle notifications for the `#daily-theme` channel updates. This option is only available on the official Midjourney Discord server.
- **/docs**: Quickly access topics in the user guide. This option is only available on the official Midjourney Discord server.
- **/describe**: After uploading an image, the Midjourney Bot will produce four descriptive prompts. This is ideal for exploring new vocabulary and visual styles. More on this in *Chapter 6*.
- **/faq**: Quickly access frequently asked questions about prompt crafting. This option is only available on the official Midjourney Discord server.
- **/fast**: Activate *Fast* mode. This is the default mode and can also be configured using the **/settings** command. It allows for faster image generation, but be mindful of your *fast hours* limits based on your subscribed plan.
- **/help**: View essential tips and information about the Midjourney Bot.
- **/imagine**: Produce an image from a given prompt.
- **/info**: Review details about your account, subscription, ongoing tasks, and much more.
- **/stealth**: Exclusive for *Pro* plan subscribers, *Stealth* mode keeps your images off `midjourney.com`. Use **/stealth** to activate and **/public** to deactivate.

  Note: While this hides images on Midjourney, images in public Discord channels remain visible. For full privacy, use **Direct Messages** or a private server (More on this in the previous chapter).
- **/public**: This activates **Public** mode and can also be configured using the **/settings** command. Disabling this option is only possible for *Pro* plan members.
- **/subscribe**: Create a personal link directing to a user's account page.
- **/settings**: Access and modify the settings of the Midjourney Bot. This is where you can choose the Midjourney model you want to use.
- **/prefer option**: Define or modify a custom option.
- **/prefer option list**: Review your set custom options.
- **/prefer suffix**: Designate a suffix for all prompts.

  *For settings and custom preferences in a more advanced mode, see *Chapter 7*.
- **/prefer remix**: Engage or disengage *Remix* mode. More on this in *Chapter 7*.
- **/relax**: Activate *Relax* mode; this mode can also be configured using the **/settings** command. Available for *Standard*, *Pro*, and *Mega* subscribers, this mode allows unlimited image creation without consuming GPU time. However, images get queued based on system usage, typically

resulting in a 0 to 10-minute wait. Your wait time can vary based on your recent *Relax* mode activity. Priorities reset monthly upon subscription renewal.

- **/show**: Use an image's **Job ID** to revive or modify a job in Discord. This command, paired with a unique Job ID, lets you transfer a job to different servers or channels, retrieve lost jobs, or update older jobs for new variations, upscaling, or incorporate the latest features. It's restricted to your jobs.

> **Job ID**
>
> Every Midjourney-generated image has a distinct Job ID.
>
> Here's an example of a Job ID's typical appearance: `9333dcd0-681e-4840-a29c-801e502ae424`.
>
> You'll find these IDs at the beginning of image filenames, within website URLs, or directly in the image's filename.
>
> **Finding the Job ID**:
>
> Go to the Midjourney site and, in your member gallery, choose **...** | **Copy...** | **Job ID**.
>
> - **Via URL**: Job IDs sit at the end of the gallery image URL – for example, `https://www.midjourney.com/app/users/23494383772740034/?jobId=8 aad7451a-3d6d-4bfb-8ad8-5529777be04a`.
>
> - **From the filename**: They appear as the last segment of an image's filename – for example, `User_Sailboats_on_the_horizon_aad7451a-3d6d-4bfb-8ad8-5529777be04a.png`.
>
> - **Via Discord emoji reaction** (More about emoji reaction will be covered in *Chapter 9* in the *Emojis for reactions* callout box):
>
> Reacting with the envelope emoji (✉) forwards a completed job to your direct messages, providing the image's seed number and Job ID.
>
> Note: The ✉ reaction is limited to your jobs:
>
> Figure 3.16 – Example of how to find the Job ID on your Midjourney page

- **/shorten**: This command optimizes and reviews your prompt. More on this in *Chapter 6* in *The right words* section.
- **/Turbo**: This engages *Turbo* mode. This mode is exclusive to Midjourney model versions 5, 5.1, and 5.2, and can also be activated via the **/settings** command. Leveraging an advanced GPU pool, it produces results up to four times quicker but uses double the GPU minutes compared to *Fast* mode.

For more detailed information about GPU hours and the *Fast*, *Relax*, *Turbo*, and *Stealth* modes, visit your Midjourney account page at `https://www.midjourney.com/account/`.

By navigating the Midjourney Bot's commands, we've unlocked a more nuanced way to interact with the platform. Each command offers specialized functionalities, enhancing our image-creating journey. With this foundation set, we're now geared up to explore the unique traits of the early Midjourney models, V1 and V2, in the upcoming section. It's all about understanding the evolution of visual capabilities.

Taking a journey through Midjourney's evolution

The beauty of any evolutionary journey, especially in technology, is seeing how the roots set the stage for today's advancements. Midjourney, today's digital powerhouse with legions of Discord fans, wasn't born as the polished platform we know now it to be. Its humbler beginnings in the early days of 2022 were full of excitement, discovery, and a bit of that rough-around-the-edges charm.

V1 and V2

When Midjourney V1 was launched in February 2022, it was like watching the first steps of a toddler – fascinating and full of potential. We all stood amazed, even if what was being produced was basic. By the time V2 rolled around in April, the toddler had begun to jog, showing off a few more tricks, revealing its potential even more clearly.

In this section, we'll walk down memory lane, reminiscing about Midjourney's early versions, V1 and V2. We'll understand not only their functionalities but also their significance in the broader journey of Midjourney. As you soak in the knowledge of these initial versions, it'll be an enlightening trip to see where it all began and how far we've come.

So, grab your nostalgic glasses, and let's journey back to those initial days when *quality* was a simple astonishment and when every new feature was a milestone. Welcome to the visual story of Midjourney's V1 and V2.

The beginning

To ensure a consistent comparison across all model iterations, we'll employ the same text prompt we used previously: **/imagine prompt** `Polaroid Photo, Golden retriever playing fetch on a sunlit beach with gentle waves, Silhouette lighting.`

Now, given that Midjourney, by default, always taps into its latest model (which, at present, is version 5.2), special steps must be taken to revisit earlier versions such as the inaugural version, V1:

- **Prompt-specific versioning**: This method entails manually appending the `--v 1` suffix to our prompt – for instance, **/imagine prompt** `Polaroid Photo, Golden retriever playing fetch on a sunlit beach with gentle waves, Silhouette lighting --v 1`. This targeted approach ensures that only this specific prompt will utilize the capabilities of V1. It's an invaluable technique for cases where the distinctive attributes of a particular version are desired. The `--v` suffix, followed by a number, allows you to select any available version for image generation, including `--v 1`, `--v 2`, `--v 4`, `--v 5`, `--v 5.1`, and `--v 5.2`.

> **Important note**
> It is crucial to enter the version suffixes (`--v 1`, `--v 2`, `--v 4`, `--v 5`, `--v 5.1`, and `--v 5.2`) with precision. Any deviation in spelling, spacing, or punctuation from the specified format may result in the system failing to recognize the command, thereby preventing image generation. Ensuring accuracy in these commands is essential for achieving the desired outcomes with the specified versions.

- **Global version setting**: Alternatively, by invoking the **/settings** command and selecting the desired version from the drop-down list, you'll set a global preference. This means that every subsequent prompt you craft will be generated using this specified model version. However, always bear in mind that each time you wish to switch to a different version, you'll need to revisit the settings and make the appropriate choice. This method is apt for those who plan on generating multiple images within the confines of a single version.

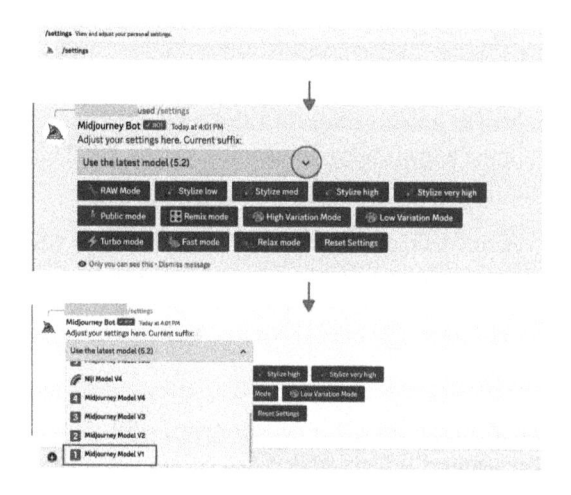

Figure 3.17 – How to change the model version using /settings command

For our demonstration, we'll begin by altering the model version using the **/settings** command. After that, we'll input the prompt in the message box and submit. This is the result:

Figure 3.18 – Using model version V1

Now, we'll navigate back to **/settings** and switch the model version to V2. Back to the message box, we'll input the same prompt and submit it. Here is the result:

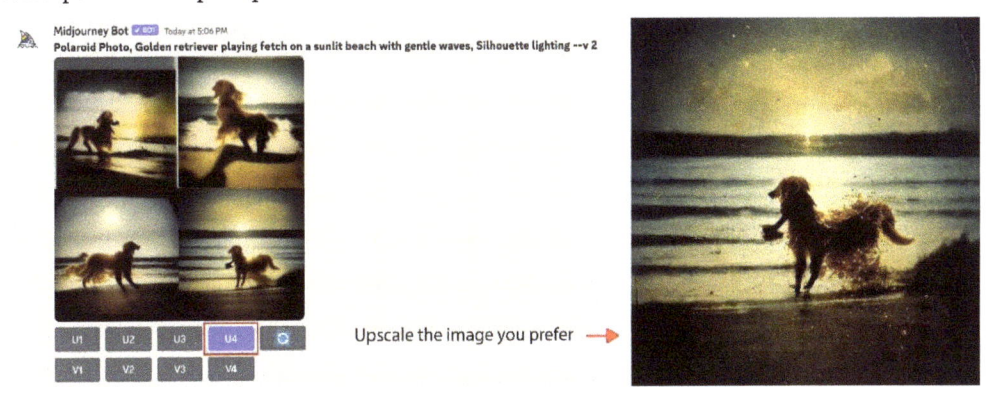

Figure 3.19 – Using model version V2

As you can see, we swiftly obtained four results for each model version. While these outputs might not seem as impressive as our initial tests with the V5.2 model, the evident progression and distinction between the two versions is noteworthy.

In these earlier versions, the images presented in the initial four-image grid were consistently generated in low resolution (256 x 256 pixels) and with fewer artifacts. Upscaling the chosen image in these versions takes more time than in newer ones. Additionally, there's a discernible disparity between the image previewed in the grid and the final upscaled result.

When examining the upscaled images (in my case, I'm going to work on the upscaled images from model version 2), you'll see options underneath that are different from what we saw when we had the V5.2 model active: We don't have the **Panning** or **Zoom** options, but we have **Make Variations** and **Vary (Region)** that we talked about before, **Upscale to Max**, **Light Upscale Redo**, **Beta Upscale Redo**, and **Remaster**:

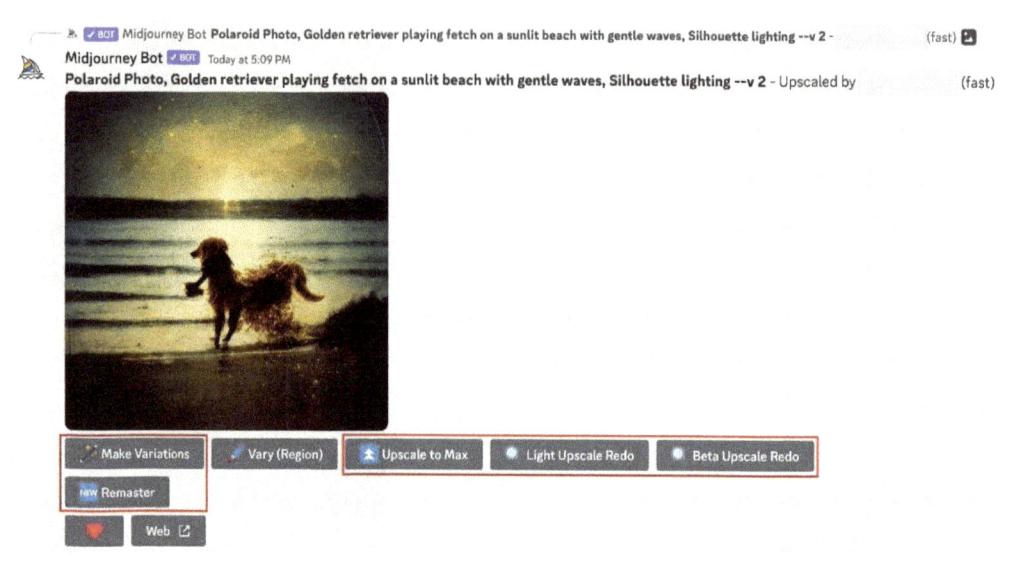

Figure 3.20 – New options after upscaling in older model versions

Let's explore the functionality of these additional variations and upscaler buttons:

- **Make Variations**: Similar to the **V1**, **V2**, **V3**, and **V4** buttons, this button generates variations of your selected image, keeping the same style and composition intact. Clicking on this button will present a new grid of four images.

- **Upscale to Max**: Max upscale is an older upscaling tool that's exclusively accessible in *Fast* mode and only for model versions 1, 2, and 3. It upscales the image to 1664 x 1664 pixel resolution.

- **Light Upscale Redo** (if adding to your prompt, use `--uplight`): Light Upscale is the smaller upscaler; it produces images with a crisp resolution of 1024 x 1024 pixels, but it provides a harmonious touch to images, perfectly calibrating them to avoid undue intensity or clutter, especially when refining portraits and sleek surfaces in earlier iterations of Midjourney.

- **Beta Upscale Redo** (if adding to your prompt, use `--upbeta`): The Beta Upscaler, for the previous model versions (V1 to V3), operates at a resolution of 1024 x 1024 pixels; for all the other model versions, it delivers a sharper 2048 x 2048 pixels output. Without cluttering images with undue details, it accentuates inherent qualities, making it especially suitable for refining portraits and smooth surfaces. With its focus on enhancing resolution, **Beta Upscale Redo** guarantees crisp and adaptable images and ensures an additional dimension of depth and realism.

- **Remaster**: The **Remaster** button creates a new grid of four images using V5.1. It's a perfect option for upscaling and refining old images.

Figure 3.21 – Images generated with each upscaler

When performing any of the previous upscalers (**Upscale to Max**, **Light Upscale Redo**, and **Beta Upscale Redo**) except for **Remaster**, a new option will appear below the generated image: **Detailed Upscale Redo**. This feature is available for model versions 1, 2, and 3 and produces a 1024 x 1024 pixels image, enriching it with additional details. In the case of our prompt, it only makes the image more intricate, but it can be a good option for adding details and textures to other types of compositions.

To see a complete list of the Midjourney dimensions and sizes for the upscalers, visit `https://docs.midjourney.com/legacy/docs/upscalers`.

Figure 3.22 – Image generated with Detailed Upscale Redo

As we rewind through Midjourney's timeline, it becomes abundantly clear how instrumental V1 and V2 were in charting the platform's ascent. These initial versions were not just about their features, but the promise they held for future innovations. With an understanding of Midjourney's foundational stages, you've gained insight into the myriad tools and features that once defined this new AI digital frontier. As we transition, our focus will shift to the enhanced capabilities and unique attributes of versions V3 and V4. Brace yourself for an enriching continuation of this technological journey, where we'll explore the maturing phases of Midjourney.

V3 and V4

The story of Midjourney's progression is nothing short of captivating. As we transitioned into V3, the platform underwent transformative changes that redefined the boundaries of AI-powered visual creation. The introduction of V3 in August 2022 brought with it sharper, more vibrant images and dynamic features such as **stylize** and **quality**. These tools granted users an unprecedented level of control, allowing for both detailed precision and AI-infused creativity.

With V4, the journey took another monumental stride forward. Building on the foundations of its predecessor, V4 fine-tuned the synergy between artistry and technology, pushing the Midjourney experience to newer heights and setting the stage for what was yet to come.

V3 – a new era of AI-driven artistry

Introduced in the heart of 2022, V3 wasn't merely an upgrade; it was a transformative force in the world of AI-based visual creation. The imagery became sharper, more pronounced, and held a certain vibrancy that we hadn't seen before. This wasn't just a technical improvement – it was a leap into a new horizon where the AI wasn't merely following instructions but adding a touch of its own artistic flair.

With the introduction of V3, light upscale's capabilities were heightened, bringing forth images that were crisp and significantly free from distortions. An essential feature – *stylize* – emerged, allowing users to decide how much they wanted the AI to add its interpretation to the images. The power was literally in the user's hands – they could decide if they wanted the AI to strictly adhere to instructions or let it wander a bit and surprise them with its own touch. Additionally, the *quality* parameter was an insight into the AI's efficiency – letting users determine the depth, clarity, and even the cost of their creations.

When we lay our eyes on the sample images generated, the evolution is unmistakable. The faces are not just clearer; they hold emotion, depth, and a level of detail that was previously unimaginable. The progression from V2 to V3 wasn't just a step; it was a giant leap.

To start exploring the differences between V2 and the innovation that V3 brought, let's take the same steps as we did before, to change models:

Enter **/settings** in the message box, then choose the model **Midjourney Model V3**.

This time, we're going to try a different prompt, and we're going to try generating a human face: /**imagine prompt** a beautiful brown-haired girl eating an ice cream in a magical garden. As you can see from the results, the reduction in distortion and artifacts/noise is significant:

Figure 3.23 – Grid generated using model V3

Now, let's understand the two new parameters, `--stylize` and `--quality`, and how can we use them to enhance our AI generations.

Understanding the --stylize parameter

One of the standout features introduced with Midjourney's V3 is the `--stylize` parameter, also represented as `--s`. This function serves as a bridge between the precise directives of the user and the innate artistry of the AI.

The core philosophy behind `--stylize` revolves around the AI's training, which leans toward producing images characterized by a rich tapestry of artistic color, balanced composition, and expressive forms. This parameter offers users the ability to dictate how ardently this training should manifest in the resulting images.

Every Midjourney model incorporates a *stylize* feature that can be manipulated to govern the artistic flair infused into the generated images. This feature works with a range of values from 0 to 1000, with 100 being the default value. However, in V3, the range stands from 625 up to 6000, with 2500 as the default value. Let's understand what can be achieved using these values:

- **Lower values (for example, 625)**: At the lower end of the spectrum, the AI remains anchored closely to the user's prompt. It delivers images that might be more literal and less infused with its own artistic interpretation.
- **Mid-range value (for example, 2500 (default))**: This acts as a balanced medium where the AI showcases its artistic flair but remains reasonably tethered to the user's instruction.
- **High values (for example, 20000 and 60000)**: Venturing into the higher values is like embarking on a creative rollercoaster. The AI takes the reins, drifting away from the literal prompt, often delivering unpredictable yet fascinating results.

To observe the power of `--stylize`, simply append `--stylize <value>` or its shorthand, `--s <value>`, to the end of your prompt. For instance, let's try it with our prompt by using five different values:

- **/imagine prompt** `a beautiful brown-haired girl eating an ice cream in a magical garden --s 650` (to lower the AI bot's interpretation)
- **/imagine prompt** `a beautiful brown-haired girl eating an ice cream in a magical garden --s 1250` (a less conservative option than the previous one)
- **/imagine prompt** `a beautiful brown-haired girl eating an ice cream in a magical garden --s 2500` (this is the stylize default value, so the result is what we have in *Figure 3.23*)
- **/imagine prompt** `a beautiful brown-haired girl eating an ice cream in a magical garden --s 30000` (letting Midjourney guide the way)
- **/imagine prompt** `a beautiful brown-haired girl eating an ice cream in a magical garden --s 60000` (just waiting to be surprised by Midjourney)

Figure 3.24 – Grids generated using the same prompt with different stylize values

The magic of the `--stylize` parameter lies in the balance it offers between human intention and AI artistry. Users can maintain strict control or let the AI surprise them with its interpretations. The choice, quite delightfully, remains with the creator. In this example, you'll observe the extent to which we've allowed Midjourney's AI to navigate the spectrum of stylization. Yet, as we approach its peak values, notable elements such as the ice cream mysteriously disappear.

Understanding the --quality parameter

The `--quality` (or `--q`) parameter is central to Midjourney's capabilities. It gives users control over the balance between time, cost, and detail in the generated image.

`--quality` determines the generation time of an image. A higher quality takes longer and uses more GPU minutes. Interestingly, it doesn't affect the image's resolution.

At the time of writing, the system entertains three distinct values for `--quality` – `.25`, `.5`, and the full 1:

- **Quality .25**: The go-to for swift, more abstract creations. Remarkably, it operates at a speed that's four times faster, consuming only a quarter of the GPU resources compared to the default.
- **Quality .5**: An optimal middle-ground, it produces outputs twice as fast and utilizes just half the GPU resources.
- **Quality 1**: This represents Midjourney's default setting, ensuring a harmonious blend of time, cost, and detail.

To apply `--quality`, simply add `--quality <value>` or `--q <value>` to the end of your prompt.

Figure 3.25 – Grids generated using the same prompt with different quality values

V4 – a quantum leap in AI-powered creativity

In a span of 6 months, from November 2022 to May 2023, the AI community witnessed the reign of Midjourney's V4. Unlike its predecessors, V4 was a groundbreaking venture, boasting an entirely new code base combined with a revolutionary AI architecture, exclusively crafted by the Midjourney team. The expanded knowledge base paved the way for a broader spectrum of characters, intricate scenarios, or multi-character scene creations.

- Let's use the same prompt as in V3, **/imagine prompt** `a beautiful brown-haired girl eating an ice cream in a magical garden`, change our setting to **Midjourney Model V4**, and generate a new mind-blowing result:

Figure 3.26 – Grid generated using the same prompt in model V4

As we can see, the evolution of this model is extraordinary; the detail and interpretation of the prompt are remarkable. In my case, I finally have a result that looks like a girl eating an ice cream in what looks like a magic garden.

Let's summarize V4's prime features:

- **Comprehensive knowledge base**: V4's grasp of subjects, creatures, environments, and tangible entities surpassed any of its predecessors. This heightened awareness guarantees the generation of visuals that resonate with reality, bridging the gap between AI-generated and human-perceived art.
- **Enhanced coherency**: One of the hallmarks of V4 was its unrivaled coherency. It showcased unparalleled finesse, especially when working with image prompts, resulting in creations that speak a thousand words.
- **Detail mastery**: This version became the benchmark for attending to minute details. Complex prompts with an intricate web of characters and objects were handled with precision, showcasing V4's might.

Midjourney V4 offers three **style** options: 4a, 4b, and 4c. These styles are fine-tuned versions of the base V4 model, trained for a particular aesthetic, bringing stylistic variations to the generated images:

- **Style 4a** is closer to the previous V4 style (from the original V4 release). It tends to produce images that are more realistic and detailed, with a focus on accuracy (prompt to input: **/imagine prompt** `a beautiful brown-haired girl eating an ice cream in a magical garden --style 4a`).

- **Style 4b** is a more experimental style that produces images that are more creative and stylized. It can sometimes produce less realistic images, but it can also produce some truly stunning and unique images (prompt to input: /**imagine prompt** `a beautiful brown-haired girl eating an ice cream in a magical garden --style 4b`).
- **Style 4c** is the default style in Midjourney V4. It is a balance between style and realism, and it is generally a good choice for most prompts (*Figure 3.26*).

--style 4a   --style 4b

Figure 3.27 – Grids generated using the same prompt in --style 4a and --style 4b

The wonders of V3 and V4 might leave many in awe, but the horizon is teeming with more technological marvels. With the latest models V5 to V5.2, AI innovation has become even more pronounced. So, as you soak in the achievements of V3 and V4, brace yourself for deeper insights and more expansive vistas in the world of Midjourney. The adventure is far from over.

V5 to V5.2 (the current model)

Journeying through Midjourney's technological timeline, we've observed progress, innovation, and leaps of genius. But on January 20, 2023, the world witnessed not just an improvement but a revolutionary step – the launch of Midjourney V5.

Imagine an artist, previously equipped with primary colors, suddenly given a palette of infinite hues. That's the scale of evolution we're discussing with V5. Not only did this version enhance the very core functionalities of the platform, but it also expanded its horizons, bridging gaps between vision and virtual reality.

To clearly understand the differences between V5 and previous versions and observe the upgrade in image generation quality in this version, let's use the prompt we used for V3 and V4, but now with the latest model.

Taking a journey through Midjourney's evolution   61

- Like before, go back to /**settings** and choose the **Midjourney Model V5.0** option; then, type the following prompt into the message box: /**imagine prompt** `a beautiful brown-haired girl eating an ice cream in a magical garden.`

As usual, a grid of four images will be generated; I've chosen to upscale the fourth image (**U4**).

As with some older model versions, the **Remaster** option appears in the buttons below the image (we previously found that in the case of model V2, **Remaster** generated an upscale to V5.1). In this case, **Remaster** upscales to the latest model version. Let's click and witness the automatic upgrade of this image to one that you'd obtain with the current V5.2:

Figure 3.28 – Grid generated using the same prompt in V5.0 and the final grid in ^V5.2 using the Remaster feature

Let's take a deeper look into the V5 series versions.

The fantastic V5.2 model

Released in June 2023, V5.2 is the pinnacle of Midjourney's innovation. A simple addition of `--v 5.2` to your prompt unlocks a universe of **hyper-realistic** visuals. This model is distinguished by its detailed, sharp outputs with impeccable colors, contrast, and compositions. Its refined understanding of prompts and heightened responsiveness to the `--stylize` parameter (values between `0` to `1000`, with the default being `100` – for example, `--stylize 750`) set it apart. The `--style raw` parameter also offers a variant that tones down Midjourney's innate aesthetic.

> **The Style Raw parameter**
>
> This parameter can be used to reduce the influence of Midjourney's default aesthetic on the generated images. This can be helpful for users who want more control over the appearance of their images, or who want to create images that are more realistic or photographic in style. In my case, adding this to the end of my prompt generates images that are closer to what I asked for: the girl is more childlike, and the background has a more magical touch.

--v 5.2   --v 5.2 --style raw

Figure 3.29 – A beautiful brown-haired girl eating an ice cream in a magical garden in --v 5.2 versus a beautiful brown-haired girl eating an ice cream in a magical garden --v 5.2 --style raw

Journeying back to V5.1

May 4, 2023, saw the release of the V5.1 model. This model carved a niche with its robust default aesthetic; it's a great choice to create a fantasy image or a piece of abstract art. Its high coherency, precision in understanding natural language, minimized artifacts, amplified sharpness, and special features such as `--tile` (more on this later) for repeating patterns made it a crowd favorite.

--v 5.1   --v 5.1 --tile

Figure 3.30 – A beautiful brown-haired girl eating an ice cream in a magical garden in --v 5.1 versus a beautiful brown-haired girl eating an ice cream in a magical garden in --v 5.1 --tile

The pioneer – the V5.0 model

V5.0, while older, has its own charm, and it's perfect for prompts that require accuracy or realism. Launched as the default from March 30 to May 3, 2023, it mastered the art of interpreting the prompt in a more literal and approximate way than previous versions, and if you are trying to create an image of a specific object or person, model version 5.0 is a good choice.

But as enthralling as V5's universe is, the Midjourney Odyssey has more wonders in store. Up next, we'll turn the pages to the colorful and imaginative world of anime as we delve deep into **Niji** versions 4 and 5. Buckle up as we set forth, exploring the vibrant nuances and transformative power of these distinct models.

Niji versions – the anime world

In the ever-evolving narrative of Midjourney's cutting-edge developments, there arises a shimmering chapter dominated by vibrant splashes of color, electric vibes, and fantastical storytelling – the **Niji models**, designed exclusively for the world of anime and cartoons.

Crafted to cater to the booming demand for authentic anime and cartoon visuals, Niji versions 4 and 5 stand as Midjourney's forefront in this domain, blending technical precision with unique illustrative styles. To get the most out of these groundbreaking models, consider the following guidelines. Here are some tips for creating good text prompts for Niji models 4 and 5:

- **Be specific**: The more specific you are in your prompt, the more likely Niji 4 is to generate an image that matches your vision – for example, specify the color of the hair, whether the setting or background is indoors or outdoors, if it's a busy street, and so on.
- **Use keywords**: When you are creating your prompt, use keywords that describe the image you want to create. For example, if you want to create an image of an anime battle scene, you could use keywords such as *action*, *fight*, *warrior*, *snow*, and *mountain*.
- **Be creative**: Don't be afraid to get creative with your prompts. The more creative you are, the more interesting the images you will generate. Although it's a specific style, why not explore different themes and concepts to understand the possibilities, such as the anime version of Superman, or what a painting such as the Mona Lisa would look like in the Niji style?

Niji V4 – the forerunner

Launched in November 2022, the genesis of Niji models was marked with this version, which has been meticulously trained to render anime and illustrative styles. Its inherent strengths lie in crafting dynamic action shots and zooming into character-centric compositions. However, its approach leans more toward the illustrative than the realistic, making it a perfect choice for action sequences and character portraits. **Niji V4** is your ally for dynamic, action-packed shots or character-driven compositions.

Next, we will experiment with the potential of the Niji V4 model by selecting it in our settings. We'll complete a new exercise by typing our previous prompt, after which we'll try a new prompt that will showcase the version's ability quite well.

Previous prompt: **/imagine prompt** `a beautiful brown-haired girl eating an ice cream in a magical garden.`

New prompt: **/imagine prompt** `A dynamic anime battle scene between two powerful warriors, one with long flowing hair and piercing blue eyes, and the other with short spiky hair and fiery red eyes.`

As you can see, the new prompt is more specific than the previous one, and it uses keywords that describe the image you want to create. It is also more creative as it introduces a new element of conflict between the two warriors:

Figure 3.31 – Grids generated with each of the two prompts with Niji V4

Niji V5 – the masterstroke

Released in May 2023, this rendition takes everything its predecessor did and amplifies it. **Niji V5**, a harmonious collaboration between Midjourney and Spellbrush, is laser-focused on manifesting realistic, detailed, and even more stylized visuals. With the potential to produce images with rich intricacy, its prowess does come at the cost of a longer generation time. But for many, the wait is well worth the reward.

Niji V5 is the go-to for intricate, stylized, and hyper-realistic portrayals, especially when the setting converges between anime and the real world.

As we did before, let's try out our prompt of the girl eating ice cream and the one more suited to this style of Niji. But first, don't forget to change the model version in your **/settings** to **Niji Model V5**.

Previous prompt: **/imagine prompt** `a beautiful brown-haired girl eating an ice cream in a magical garden.`

New prompt: **/imagine prompt** `A dynamic anime battle scene between two powerful warriors, one with long flowing hair and piercing blue eyes, and the other with short spiky hair and fiery red eyes:`

Figure 3.32 – Grids generated with each of the two prompts with Niji V5

Beyond the base models, Niji V5 introduces a myriad of style parameters, allowing us to customize and refine our outputs:

Figure 3.33 – You can select the styles in the settings or by adding --style cute, --style scenic, --style original, or –style expressive to your prompt

- **Original Style** (if adding to your prompt, use `--style original`): A nod to its roots, replicating the version available before May 26, 2023. If you are trying to create an image that is unlike anything you have seen before, **Original Style** is a good choice:

Figure 3.34 – Using --style original

- **Cute Style** (if adding to your prompt, use `--style cute`): For when you crave that extra dash of adorability. It's the perfect choice to create an image of a cute animal or a heartwarming scene:

Figure 3.35 – Using –style cute

- **Expressive Style** (if adding to your prompt, use `-style expressive`): Where emotions take center stage, this is perfect for ideas that require sophistication or emotions:

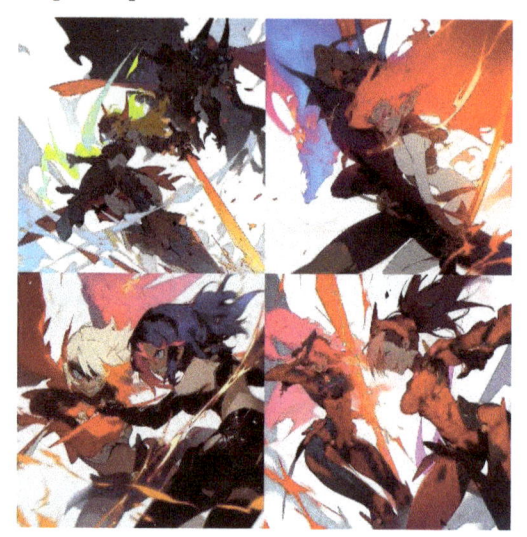

Figure 3.36 – Using --style expressive

- **Scenic Style** (if adding to your prompt, use `--style scenic`): This is perfect for landscapes and wide-angle shots. This style is designed to create beautiful backgrounds and cinematic character moments in the context of their fantastical surroundings. It is a good choice for users who want to create images that are both visually stunning and emotionally evocative:

Figure 3.37 – Using --style scenic

In conclusion, the Niji models by Midjourney are powerful tools for creating anime and cartoon illustrations. Niji V4 captures the essence of anime, while Niji V5 offers a wider range of styles and features. Both versions offer unique strengths, catering to a wide range of artistic desires. Creators can use Niji V5 to paint a broader range of stories and emotions. With these tools, the possibilities for anime artistry seem endless, promising a future where our imagination is the only limit.

Summary

This chapter provided a deep dive into the world of crafting meaningful prompts with *Imagine and prompt*, where we learned that the structure and details within a prompt are paramount in ensuring the AI produces an image closely aligned with our vision.

In *My first prompt – now what?*, we navigated the steps following your initial prompt creation, demystifying the process of the Midjourney console interpretation.

Progressing further, *Beyond /imagine – command list* offered a detailed guide on a myriad of commands, ensuring you're well-equipped to get the most out of your image generation endeavors. We retraced Midjourney's development journey through *Model versions 1 and 2*, learning about their foundational stages, and then moved forward to uncover the enhanced capabilities of *Model versions 3 and 4*.

With technological advancements in full swing, *Model versions 5 to 5.2* highlighted the latest and most innovative features, underscoring Midjourney's commitment to growth and improvement. Rounding off our journey, *Niji versions 4 and 5* introduced us to the colorful and dynamic domain of anime and cartoons, showcasing a beautiful blend of technical excellence and vivid imagination.

The insights and skills you've gained in this chapter will empower you to interact with Midjourney's offerings more effectively, enabling you to generate images that align with your vision.

Looking ahead, our next chapter will build on this foundation, introducing more advanced techniques and strategies, ensuring a logical and enriching continuation from where we've left off.

4
Understanding and Learning Parameters

In the ever-evolving world of AI art creation, understanding and using **parameters** in Midjourney is essential for creating artworks that closely match your vision. This chapter will demystify parameters and provide you with the knowledge and skills you need to master them.

We will begin with a clear explanation of what parameters are and how they affect the output of Midjourney. Next, we will provide a detailed list of available parameters, paired with examples to illustrate their functions. Finally, we will discuss legacy parameters and where to use them.

In this chapter, we will cover the following topics:

- **What are parameters?**: Here, we will introduce the fundamental concept of parameters in Midjourney, explaining their crucial role in refining the AI's creative output. This section serves as the foundation for understanding how parameters can be manipulated to achieve more precise and controlled artistic results.
- **List of basic parameters and examples**: Readers will get a comprehensive exploration of basic parameters, complete with examples to illustrate how each one influences the generation of AI art. This segment is designed to familiarize you with the tools necessary for varying and enhancing your creative outcomes.
- **List of legacy parameters and examples**: This part revisits the features of earlier Midjourney versions, providing insights into their continued relevance and application. Through examples, we illuminate how these parameters can be used to resurrect the distinctive styles and effects of past models.

In short, by the end of this chapter, you will be empowered to take your AI art creations to the next level. Welcome to a chapter that promises a deeper and more controlled exploration of the Midjourney tool.

Technical requirements

To make the most of this chapter and follow along with the hands-on activities, you'll need to ensure you've set up everything we covered in *Chapter 2*. Here's a quick recap:

- **Midjourney account**: You'll need an active account on Midjourney. If you haven't registered yet, refer back to *Chapter 2* for a step-by-step guide.
- **Discord installation**: Ensure you have Discord installed on your device since we'll be working with it alongside Midjourney. Again, the installation guide is available in *Chapter 2*.
- **Payment plans**: At the time of writing, Midjourney has no free or trial plan. A Midjourney paid plan is necessary. Please ensure you've selected the most suitable plan for your needs.
- **Server creation**: Having your own server can significantly enhance and expedite your experience with Midjourney. If you're yet to set up a server in Discord, *Chapter 2* covers the process in detail.

What are parameters?

Parameters in Midjourney are settings that allow you to have more control over the AI's creative process. They can be used to fine-tune the style, composition, and overall look and feel of your images. By understanding and using parameters effectively, you can achieve a wider range of creative results and produce artwork that more closely matches your vision.

There are two main types of parameters in Midjourney:

- **Basic parameters**
- **Legacy parameters**

In this section, we will show you how to use parameters in Midjourney and explain how to use them to create different types of AI art. We will also provide some tips to achieve more specific results. It is important to note that parameters are always added to the end of a prompt, and they are declared using double hyphens (`--`).

For example, to generate a landscape image with a different aspect ratio than the default `1:1`, you should enter the `--ar` flag at the end of your prompt. Let's try an example: **/imagine prompt** `cute robot walking in a busy NY city --ar 16:9`:

prompt The prompt to imagine

/imagine prompt `cute robot walking in a busy NY city --ar 16:9`

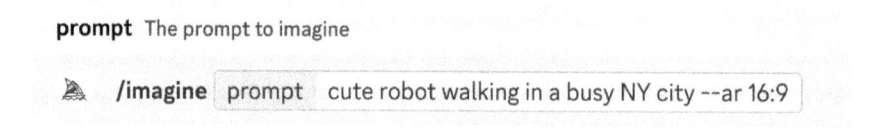

Figure 4.1 – Always use parameters at the end of your prompt

This prompt will generate a wide-angle view image with an aspect ratio of `16:9`. This is a common aspect ratio for computer monitors and television screens. In case you're wondering, here's the result:

Figure 4.2 – Using aspect ratio parameter

Another approach we can explore is the use of multiple parameters in our prompt. Let's experiment by utilizing the same prompt, but this time, in addition to the aspect ratio, we will also infuse a style. Recalling what we learned in the previous chapter, we will integrate the parameter from the Niji 5 version, `--style cute`, which could offer intriguing possibilities for the type of prompt we are examining. In summary, our refined prompt will be structured as follows: **/imagine prompt** `cute robot walking in a busy NY city --ar 16:9 --niji 5 --style cute`:

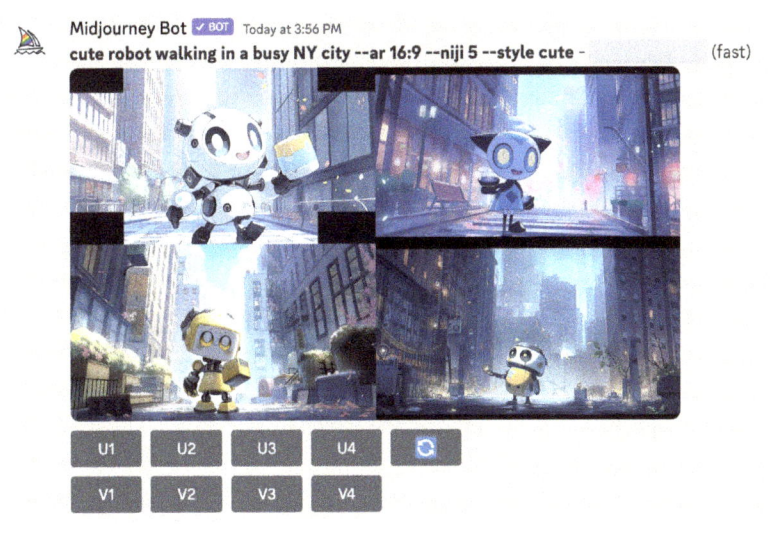

Figure 4.3 – Using aspect ratio parameter and model version styles

The new outcome is really surprising, and the style we've opted for imbues the image with very interesting characteristics.

So, now, let's explore the variety of available parameters and uncover their respective functions. Let's begin our journey of discovery, unlocking the myriad ways in which parameters can unleash your full creative potential with Midjourney.

List of parameters and examples

Midjourney parameters are like the tools of an artist, allowing you to create more detailed, controlled, and creative AI art. In this section, we will explore each parameter, explaining its purpose and demonstrating how it can be used to shape and refine your artistic vision.

Through real-life examples, we will see how each parameter alters the artistic output. This hands-on approach will give you not only theoretical knowledge but also practical insights into how to use parameters to create the AI art you want.

By the end of this section, you will have a deeper understanding of Midjourney parameters and an enriched toolkit of skills, enabling you to craft AI art with greater precision and diversity.

Let's begin by going through the list of basic parameters.

Basic parameters

Basic parameters control the overall style and composition of your image, such as aspect ratio, chaos level, and quality. Let's take a look at the list of basic parameters available to us.

Aspect ratios (`--aspect <value:value>`, `--ar <value:value>`)

Aspect ratios are a fundamental concept representing the relationship between the width and height of an image or screen. They are important in Midjourney because they determine the dimensions of the generated artwork. The default aspect ratio is `1:1`, which is a square. However, you can use the `--aspect` or `--ar` flag to change the aspect ratio of the image you want to generate.

The following are common aspect ratios:

- `1:1` – Square
- `5:4` – A common choice for frames and prints
- `3:2` – Prevalent in print photography
- `7:4` – Approximates HD TV and smartphone screens

Figure 4.4 – Example of common aspect ratios

You can choose an aspect ratio based on your artistic intentions. For example, you might choose a widescreen aspect ratio for a landscape scene or a narrow aspect ratio for a portrait. You can also use aspect ratios to evoke specific emotions or associations in the viewer. For example, a widescreen aspect ratio can create a sense of grandeur and expansiveness, while a narrow aspect ratio can create a more intimate and claustrophobic feeling.

To change the aspect ratio in Midjourney, simply add the `--aspect` or `--ar` flag to your prompt, followed by the desired aspect ratio. For example, to generate an image with an aspect ratio of 5:4, you would add the following to your prompt:

`--aspect 5:4` or `--ar 5:4`

> **Tips for using aspect ratios**
>
> Keep in mind that aspect ratios can affect the composition of your image. For example, a widescreen aspect ratio can make your image feel more expansive, while a narrow aspect ratio can make it feel more claustrophobic.
>
> If you are unsure which aspect ratio to use, start with the default 1:1 aspect ratio. This is a good all-purpose aspect ratio that will work well for most images.
>
> You can also experiment with different aspect ratios to see how they affect the look and feel of your images.

Here are two good prompt examples to use in two different aspect ratios:

- **Aspect ratio**: 1:1 (Default) – In this case, you don't need to add the `--ar 1:1` flag, as it is the default value

    /imagine prompt `A surreal portrait of a cat with big eyes, in the style of Salvador Dali --ar 1:1`

    This prompt is well suited for a 1:1 aspect ratio because it is a balanced composition. The cat will be the main focus of the image, and the big eyes will add a touch of surrealism. The 1:1 aspect ratio will also allow you to capture the cat's entire face in detail:

    Figure 4.5 – Grid of four images generated using the default aspect ratio in /imagine prompt - A surreal portrait of a cat with big eyes, in the style of Salvador Dali --ar 1:1

- **Aspect ratio**: 5:4

    /imagine prompt `A sweeping landscape of the Grand Canyon, with a river winding through it --ar 5:4`

    This prompt is well suited for a 5:4 aspect ratio because it will allow you to capture the grandeur of the Grand Canyon. The 5:4 aspect ratio is also a popular aspect ratio for prints, so your image could potentially be printed and framed:

Figure 4.6 – Grid of four images generated using /imagine prompt - A sweeping landscape of the Grand Canyon, with a river winding through it --ar 5:4

You can use these prompts as a starting point, and then experiment with different variations to see what kind of results you get. For example, you could try using different art styles, different colors, or different lighting conditions (more on this later). You can also try adding additional elements to your prompts, such as people, animals, or objects. Have fun experimenting with the *aspect ratio* parameter!

Chaos (`--chaos <value between 0 and 100>` *or* `--c <value between 0 and 100>`*)*

Midjourney's introduction of the **chaos** parameter adds a brush stroke of unpredictability and diversity to your AI-generated art. This parameter controls how varied the results will be. The *chaos* parameter accepts values between `0` and `100`, with `0` representing the lowest level of chaos and `100` representing the highest level of chaos. At a chaos value of `0` (this is the default value), Midjourney will generate images with a high degree of similarity and consistency. As the chaos value is increased, Midjourney will begin to introduce more variation into the images, resulting in a wider range of outcomes.

Here is the spectrum of outcomes with examples:

- **No chaos** (`--chaos 0` or `--c 0`)

    Similar and consistent initial grids with every job run.

78　Understanding and Learning Parameters

/**imagine prompt** `Panda smelling flowers --c 0:`

Figure 4.7 – Grid of images generated using /imagine prompt - Panda smelling flowers --c 0

- **Low chaos** (`--chaos 10` or `--c 10`)

    Slight variations, retaining reliability in outcomes.

    /**imagine prompt** `Panda smelling flowers --c 10:`

Figure 4.8 – Grid of images generated using /imagine prompt - Panda smelling flowers --c 10

- **Moderate chaos** (`--chaos 25` or `--c 25`)

    Balanced unpredictability, introducing noticeable diversity.

    /imagine prompt `Panda smelling flowers --c 25`:

    Figure 4.9 – Grid of images generated using /imagine prompt - Panda smelling flowers --c 25

- **High chaos** (`--chaos 50` or `--c 50`)

    Amplified variations, unveiling unexpected compositions.

    /imagine prompt `Panda smelling flowers --c 50`:

    Figure 4.10 – Grid of images generated using /imagine prompt - Panda smelling flowers --c 50

80   Understanding and Learning Parameters

- **Very high chaos** (`--chaos 80` or `--c 80`)

    Extreme variability, showcasing a myriad of artistic mediums and compositions.

    /imagine prompt `Panda smelling flowers --c 80`:

    Figure 4.11 – Grid of images generated using /imagine prompt - Panda smelling flowers --c 80

- **And a bonus image** (`--chaos 100` or `--c 100`)

    You can get a panda, or you may not get a panda. The best thing is just to expect the unexpected.

    /imagine prompt `Panda smelling flowers --c 100`:

    Figure 4.12 – Grid of images generated using /Imagine prompt - Panda smelling flowers --c 100

Fast (`--fast`)

Midjourney's **fast** parameter is the perfect feature for users who want to speed up the image generation process. This parameter is available to all subscribers from the *Basic* plan to the *Mega* plan and is the default mode for generating images.

To use the *fast* parameter, simply add the `--fast` flag to your prompt, as in the following example:

/imagine prompt `A cat sitting in a tree --fast`

When you use the *fast* parameter, it overrides your current setting and runs this single job using *Fast* mode. Midjourney will then generate your image using a faster algorithm. This can reduce the generation time by up to 50%.

The *fast* parameter is a good option to use when you need to generate images quickly, such as when you are working on a time-sensitive project or when you are experimenting with different ideas. However, it is important to keep in mind that the different subscription plans (`https://docs.midjourney.com/docs/plans`) have different amounts of monthly GPU time.

Image weight (`--iw <value between 0 and 2>`)

This parameter is relevant only when utilizing both image and text prompts (more on this in *Chapter 5*). The **image weight** parameter, `--iw`, sets the image prompt weight relative to the text. Adjusting the value of the `--iw` flag alters the significance of the image in the final output. The default value is `1` in model version 5 in a range of `0` to `2`. A higher value amplifies the image's impact, while a lower value lets the text take the spotlight:

- **Lower image weight** (`--iw .5`, `--iw .75`):

    A textual element, such as `pillow`, dominates, with subtle influences from the image prompt.

    /imagine prompt `cat.jpg printed on a pillow --iw .5:`

Figure 4.13 – Grid of images generated using /imagine prompt - cat.jpg printed on a pillow --iw .5

- **Higher image weight** (`--iw 1.25, --iw 1.5, --iw 1.75, --iw 2`):

    The image prompt significantly influences the theme and features of the generated output, resulting in a more visually cohesive blend with the text.

    - **Example 1**: /**imagine prompt** `cat.jpg printed on a pillow --iw 1.25`

    The higher the values, the more the image takes precedence over the text.

    - **Example 2**: /**imagine prompt** `cat.jpg printed on a pillow --iw 2`

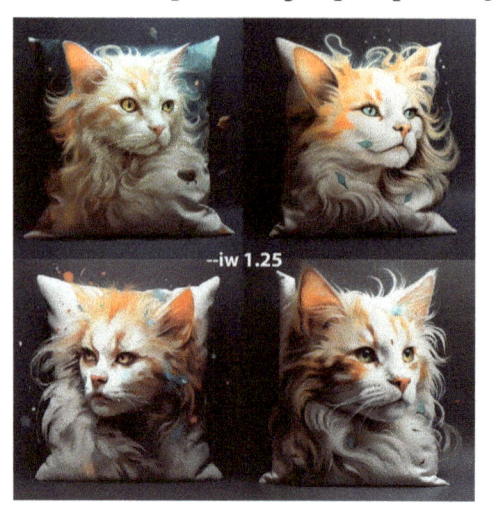

Figure 4.14 – Grids of images generated using /imagine prompt - cat.jpg printed on a pillow --iw 1.25 and /imagine prompt - cat.jpg printed on a pillow --iw 2

No (`--no`)

The **no** parameter directs the Midjourney Bot to exclude specified elements from appearing in your image. It's perfect for refining complex compositions, yielding more focused artwork.

To employ the exclusionary artistry of `--no`, simply append it to your prompt with the undesired elements. In case you want to exclude more than one element, just separate them with commas. Let's try a prompt that I know will give me elements I will ask later to exclude:

/**imagine prompt** `A cat sitting in a tree:`

Figure 4.15 – Grid of images generated using /imagine prompt - A cat sitting in a tree

As I imagined, we have lots of green natural elements. Let's experiment with the same prompt using --no to exclude everything green and then generate another version in which we exclude the leaves as well as the green:

Figure 4.16 – Grids of images generated using /imagine prompt - a cat sitting in a tree: --no green and /imagine prompt - a cat sitting in a tree --no green, leaves

84   Understanding and Learning Parameters

Quality *(*`--quality <value of either .25, .5, or 1>`*,* `--q < value of either .25, .5, or 1>`*)*

The **quality** parameter controls the quality of the generated image. Like an artist's brush controlling how much detail and texture is generated in an image, it works by modifying the amount of time spent generating the image.

Values accepted are `.25`, `.5`, and `1`, with the default set at `1`. Higher values produce more detailed images but take longer and use more GPU minutes. Lower values produce less detailed images but are faster and use fewer GPU minutes.

The *quality* parameter can be used to create a variety of effects, depending on the artist's intent. For example, lower settings can be used to capture the essence of abstract spontaneity, while higher settings can be used to accentuate intricate details in architectural visuals.

Let's take a quick look at the scales of quality:

- `--quality 1`: The default setting, investing full GPU minutes for a detailed masterpiece:

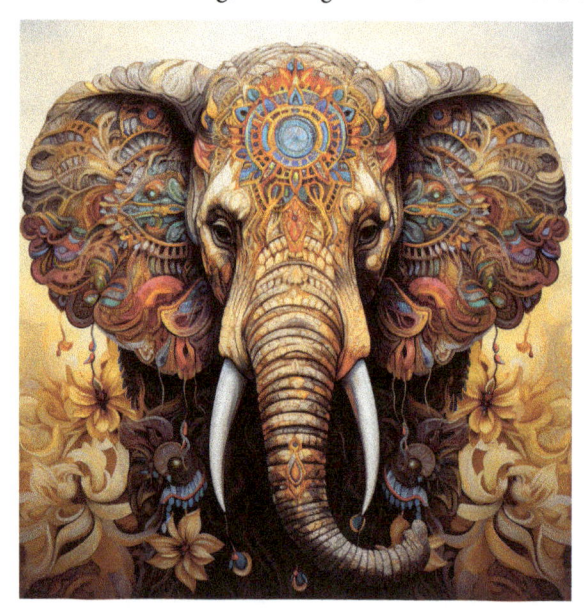

Figure 4.17 – /imagine prompt - detailed exotic elephant portrait --q 1

- `--quality .25`: Quickest, yet least detailed, utilizing a quarter of the GPU minutes:

Figure 4.18 – /imagine prompt - detailed exotic elephant portrait --q .25

- `--quality .5`: A balanced approach, crafting less detailed results in half the GPU minutes:

Figure 4.19 – /imagine prompt - detailed exotic elephant portrait --q .5

This parameter and its values are compatible with model versions 4, 5, 5.1, 5.2, and Niji 5.

Relax (`--relax`)

Midjourney offers a tranquil alternative to the *fast* parameter we talked about earlier, with the **relax** parameter. *Standard*, *Pro*, and *Mega* plan subscribers can create an unlimited number of images each month in *Relax* mode. While this mode doesn't consume any GPU time, jobs are systematically queued based on individual system usage, ensuring fairness and accessibility for all.

When you use the *relax* parameter, your current settings are overridden, and a single job is run in *Relax* mode.

To use the *relax* parameter, simply add `--relax` to the end of your prompt as in the following example:

/imagine prompt `detailed exotic elephant portrait --relax`

Repeat (`--repeat <value>` or `--r <value>`)

The **repeat** parameter allows the rendering of multiple variations of your prompt, augmenting the pace and variety of your AI art by executing a job multiple times and generating diverse sets of images per prompt. This parameter is only available in *Fast* and *Turbo* modes, and the range of values (number of times you want to repeat the prompt) is between 2 and 40, varying according to your subscription level:

- **Basic subscribers**: 2–4 times
- **Standard subscribers**: 2–10 times
- **Pro and Mega subscribers**: 2–40 times

Using my last prompt, let's add the *repeat* parameter with the value 5 `--r 5`:

/imagine prompt `detailed exotic elephant portrait --q .5 --r 5`

Next, the Midjourney Bot will ask you if you really want to re-roll the prompt five times. Click **Yes** to confirm.

A new message will appear informing you that your jobs are being processed. You will also have the following options:

- **Show Template**: View the complete prompt
- **Cancel All**: Cancel all pending jobs

To view the complete prompt, click **Show Template**:

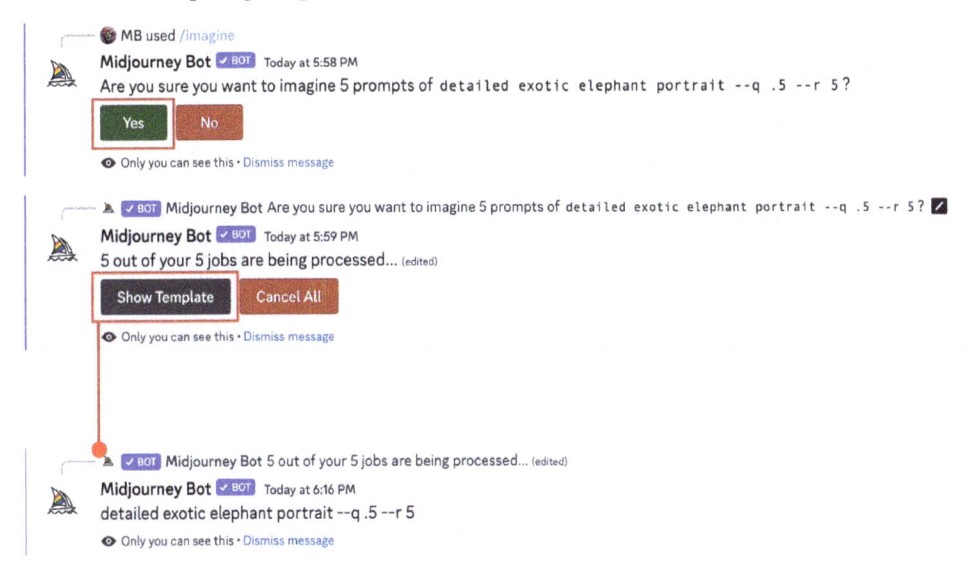

Figure 4.20 – Midjourney Bot messages for the repeat parameter

Seed (--seed <number>)

The **seed** parameter in Midjourney is used to specify the random number generated for each image. This means that if you use the same seed number and a prompt, you will get two very similar images, even if the prompts are different.

The *seed* parameter can be useful for a number of reasons. For example, you can use it to do the following:

- Recreate an image with a slightly different prompt or settings
- Generate a series of variations on an image

You can find the seed of any Midjourney image by reacting to it with the **envelope** emoji (right-click on the image you want to use as a seed, then select **Add Reaction | envelope** emoji) (*Figure 4.21*) or by sending a direct message to the Midjourney Bot (right-click on the image you want to use as a seed, then select **Apps | DM Results**) (*Figure 4.22*).

88   Understanding and Learning Parameters

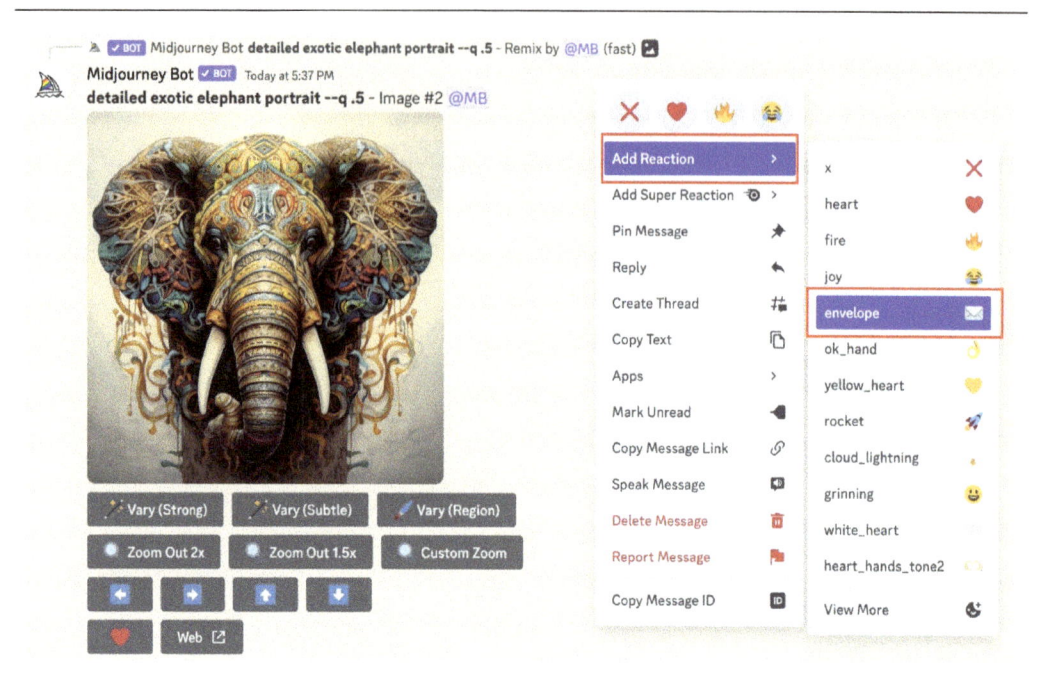

Figure 4.21 – Right-click on the image you want to use as a seed, then choose Add Reaction and envelope

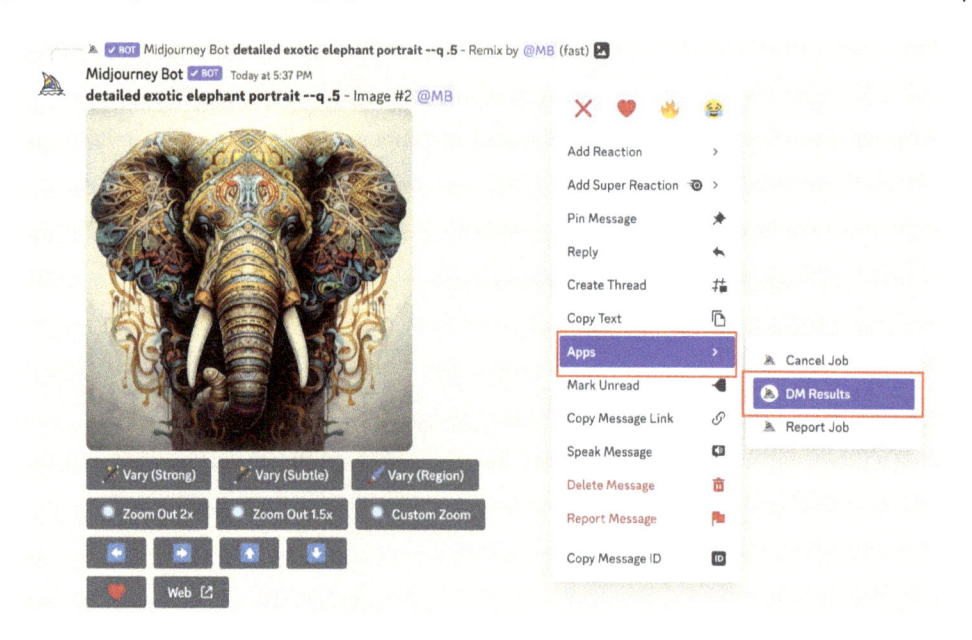

Figure 4.22 – Right-click on the image you want to use as a seed, then choose Apps and DM Results

List of parameters and examples   89

After choosing either of these two options, the Midjourney Bot will pop up on the left side of your screen next to your server's list, indicating you have a message. Click on it, and information about the job ID and seed number will be available:

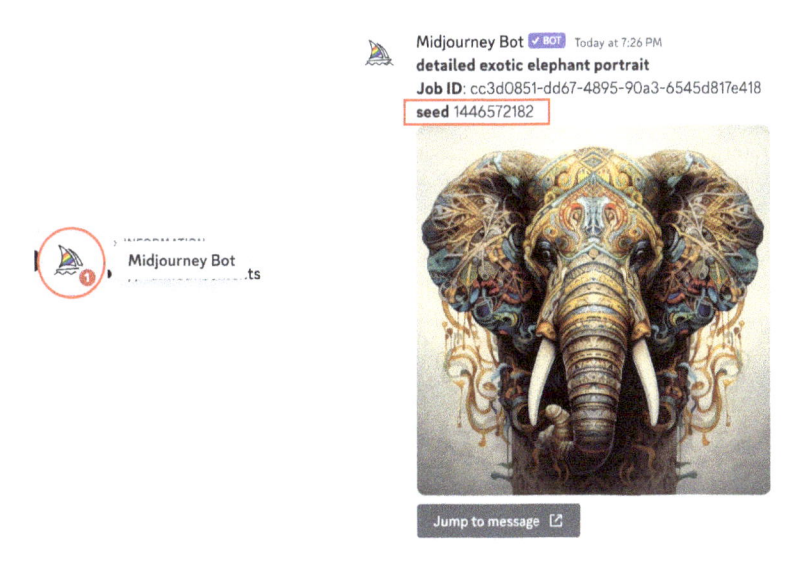

Figure 4.23 – Click on the Midjourney Bot to access the message with your seed number

Now that we have the seed number, let's use it to create a new set of images. In the following example, I've used the exact same prompt two times with the seed value, and the results are almost identical. Without the use of the *seed* parameter, this would never be possible.

Figure 4.24 – /imagine prompt - detailed exotic elephant portrait --q .5 --seed 1446572182

90 Understanding and Learning Parameters

> **Note**
> Seed numbers are dynamic and should not be relied upon between sessions, which means that whenever you log out or have an inactive Discord account, that number is no longer valid. Basically, you have to go through the process of finding the seed again.

Stop (`--stop <value between 10 and 100>`)

The **stop** parameter allows you to stop the generation process at any time and control the clarity of your image. This parameter accepts values between `10` and `100`, with lower values creating abstract, blurry results. The value represents the percentage at which the current job is terminated. For instance, `10` indicates that the job is terminated after 10% completion of the job. *Stop* can be an interesting parameter to experiment with, for those who want to obtain more ambiguous results or other types of creative interpretations.

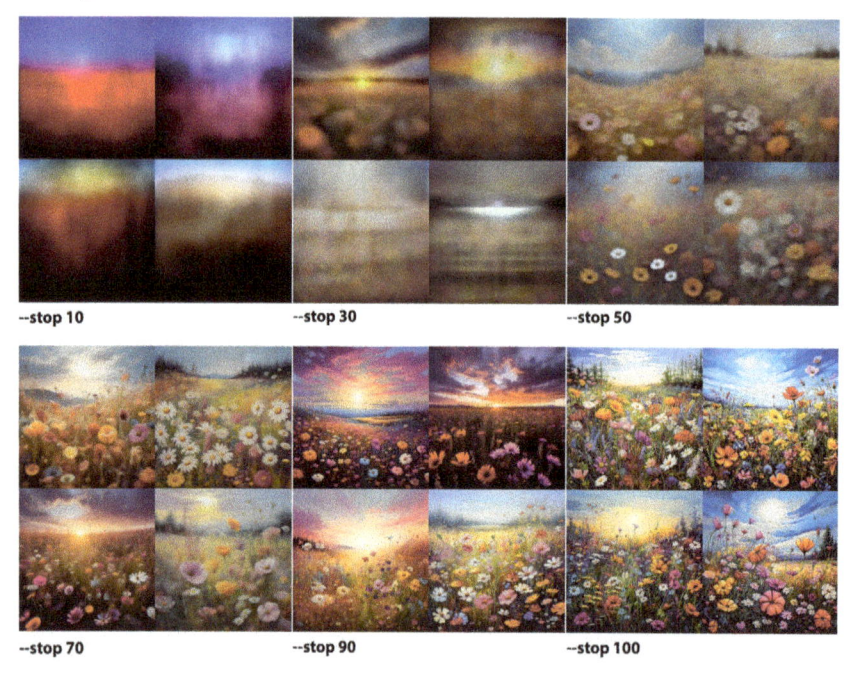

Figure 4.25 – Example of the used /imagine prompt - A field of wildflowers in a riot of colors --stop 10

Style (`--style <style>`)

The **style** parameter allows you to specify a style for the generated image. This parameter fine-tunes the aesthetic of some Midjourney model versions, helping you create more photo-realistic images, cinematic scenes, or cuter characters. As learned in our previous chapter, we have many styles we can append to our prompts:

- **Model versions 5.1 and 5.2**: `--v 5.1`, `--v 5.1 --style raw`, `--v 5.2`, `--v 5.2 --style raw`
- **Niji 5 style**: `--style cute`, `--style scenic`, `--style original`, or `--style expressive`
- **Version 4 model**: `--style 4a`, `--style 4b`, `--style 4c`

Stylize (`--stylize <value between 0 and 1000>` *or* `--s <value between 0 and 1000>`*)*

The versatile **stylize** parameter enables you to control the level of stylization, allowing you to experiment with a spectrum of details, textures, and artistic expressions. It accepts values between 0 and 1000, with 100 being the default number. These numbers dictate the strength of stylization applied to your generated images; lower values yield images closely aligned with the prompt but more simple and raw, while higher values produce distinctly artistic, albeit less accurate, representations.

Let's test this parameter with a new and delicious prompt using different values (note that unless I mention a different model version, all my prompts are always in the current default V5.2):

/imagine prompt `Donuts with icing on pastel pink background. Sweet donuts. top view assorted with various chocolate glazed and sprinkles, sugar sweets concept --stylize 0`

Next, you can see the difference between the various `stylize` values:

Figure 4.26 – Example of a grid using lower stylize values --stylize 0 and --stylize 75

Figure 4.27 – Example of a grid using default and high stylize values --stylize 100 and --stylize 250

Figure 4.28 – Example of a grid using high and very high stylize values --stylize 500 and --stylize 750

The *stylize* parameter is compatible across multiple Midjourney versions and models, including Niji 5 and V4, each offering unique stylistic outcomes; however, for V5.2, consider reducing stylization values to 20% of previous settings for equivalent outcomes, since this model version is more sensitive to different *stylize* values.

Tile (`--tile`)

The **tile** parameter generates a tiled image from the prompt. This can be useful for creating large images or for creating repeating patterns perfect for fabrics, interior decor, wallpapers, and various textures. This parameter works seamlessly with model versions 1, 2, 3, `test`, `testp`, 5, 5.1, and 5.2. Just add `--tile` to the end of your prompt and let the magic begin.

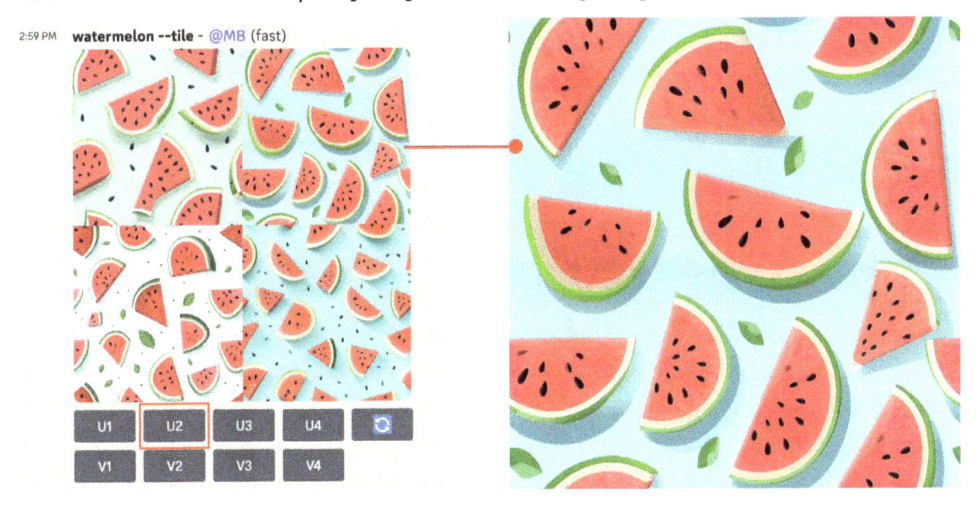

Figure 4.29 – Using /imagine prompt - watermelon --tile

To achieve a repeating pattern, you can use Photoshop or a tool such as Seamless Texture Checker (`https://www.pycheung.com/checker/`). Using Seamless Texture Checker, click the **File** button and then choose the image you want to upload from your computer:

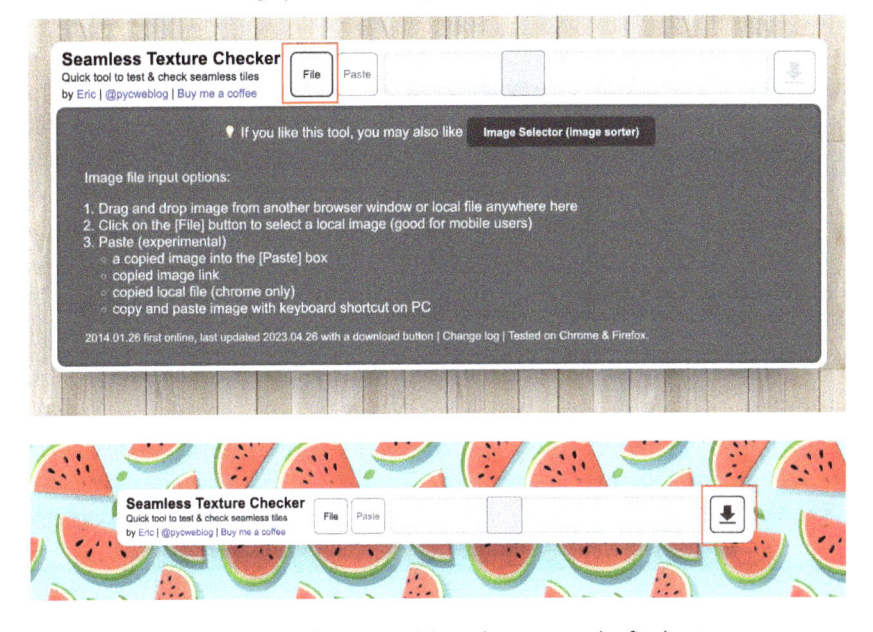

Figure 4.30 – Step-by-step guide on how to get the final pattern

The tool will almost instantaneously upload the selected. Just click on the down arrow button to download, and that's it. You have a pattern to use!

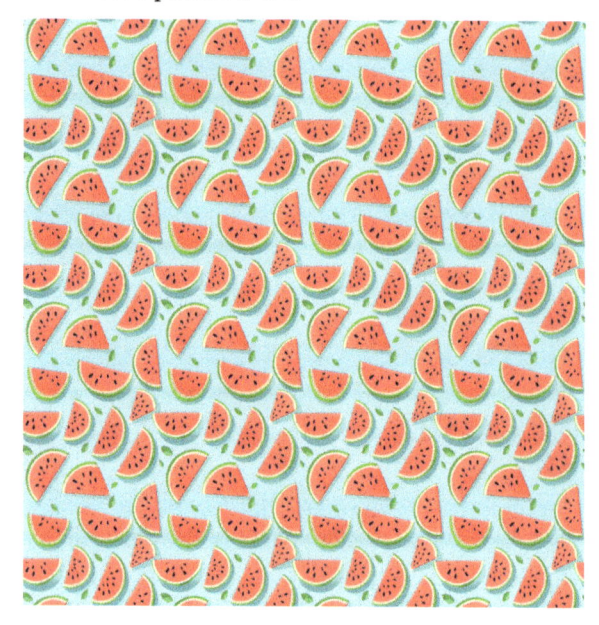

Figure 4.31 – The final pattern

Turbo (`--turbo`)

Available to *Standard*, *Pro*, and *Mega* plan subscribers, and exclusive to Midjourney model versions 5, 5.1, and 5.2, this parameter is similar to the *Fast* mode parameter, but it is even faster. However, *Turbo* mode's speed comes at a cost, consuming twice as many subscription GPU minutes as *Fast* mode. This makes it a resource-intensive option, but it can be a valuable feature for artists who need to generate images quickly. As an experimental feature, *Turbo* mode's availability and pricing structure are subject to change. When you use the **turbo** parameter, this overrides your current settings and runs this single job using *Turbo* mode.

Video (`--video`)

The **video** parameter doesn't actually generate a true video. It creates a short clip showcasing the image generation process, like an animated GIF. The length of this video isn't explicitly defined but depends on the complexity of the image being generated and the Midjourney model version used, but we can say it's around 5 to 10 seconds long. Video is exclusive to image grids, so you cannot use it for upscaling. To implement the *video* parameter, just add `--video` to the end of your prompt. After the image grid is generated, react to the finished job with the **envelope** emoji (it's the same steps we did to get the *seed* parameter previously):

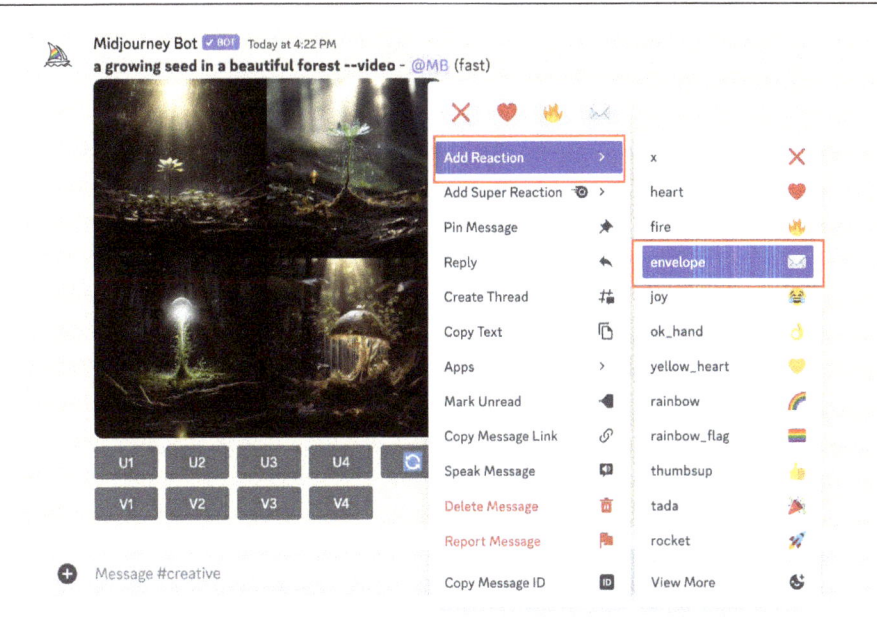

Figure 4.32 – Right-click on the grid, choose Add Reaction, and then envelope

The Midjourney Bot will pop up on the left side of your screen next to your server's list, indicating you have a message. Click on it, and information about the job ID and seed number will appear, along with your video!

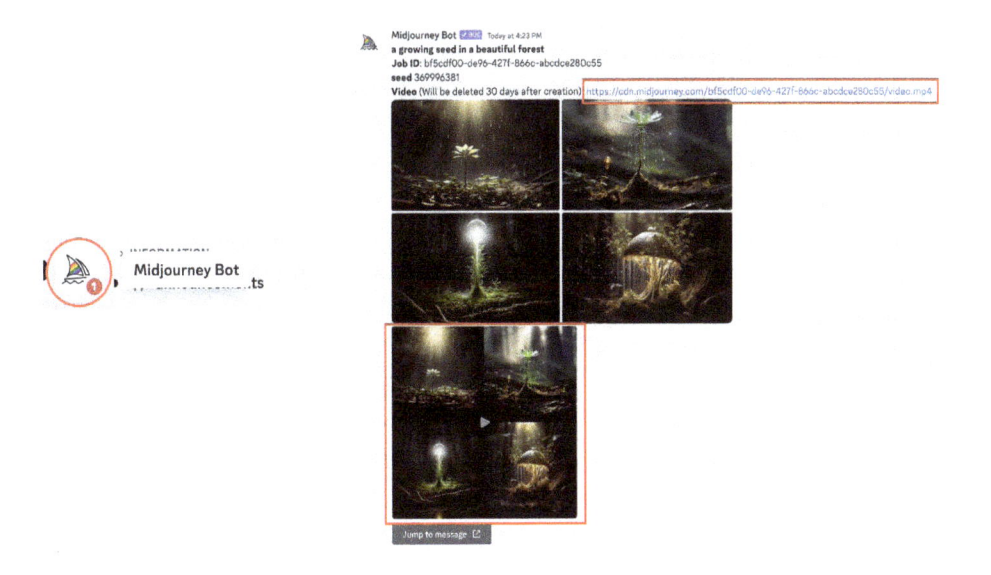

Figure 4.33 – Click on the Midjourney Bot to access the message with your video

Weird *(*`--weird <value between 0 and 3000>` ***or*** `--w <value between 0-3000>`*)*

Midjourney's **weird** parameter is powerful for artists who want to create unconventional and unexpected AI-generated images. This parameter can be used to add a touch of quirkiness, edge, or even surrealness to your work.

To use the *weird* parameter, simply add `--weird` or `--w` to your prompt. You can then specify a value from `0` (default) to `3000`. The higher the value, the more weird your image will be.

Start with a low value, such as `250` or `500`, and experiment to see what kind of results you get.

You can also use the *weird* parameter with other parameters, such as *stylize* and *chaos*, to create even more unique and unexpected images.

Keep in mind that the *weird* parameter is experimental, so you may get some unpredictable results. But that's part of the fun!

Here are a few examples of how the *weird* parameter can be used to create unexpected AI-generated images:

/**imagine prompt** `little duck dancing:`

Figure 4.34 – With --weird 0: a "standard" little duck dancing; with --weird 250: a more abstract and experimental little duck dancing

Figure 4.35 – With --weird 500 and --weird 1000: a completely surreal and otherworldly little duck dancing

I think the *weird* parameter is one of the most creative features of Midjourney, as it increases the chance of generating unexpected and unusual results. It is experimental, so there is no limit to what you can create. So, have fun and experiment!

Navigating through the basic parameters of Midjourney has empowered us to refine control over style, composition, and quality, unveiling a rich spectrum of artistic possibilities. Through exploring aspect ratios, chaos levels, and various stylistic elements, we are now prepared to explore the realm of legacy parameters, where we will meet the nuances of older models and learn to utilize their distinct capabilities to further broaden our creative horizons.

Legacy and special parameters

The term *legacy* in Midjourney refers to older versions of the Midjourney model. These versions are still available to use, but they are not as advanced as the latest version of the model, and some of the new parameters and options may not be fully compatible. Next is a list of specific parameters for older or special models. These parameters can be used to change model versions and are parameters that can be used with older versions.

Aspect ratios (`--aspect <value:value>, --ar <value:value>`)

The definition is the same as for the basic parameter we talked about before; however, for older models, there are some differences in terms of the available aspect ratios:

- **V5 and Niji 5** – Any aspect ratio
- **V4c and Niji 4** – `1:2` to `2:1`
- **V4a and 4b** – `1:1, 2:3, 3:2`

- **V3** – 5:2 to 2:5
- **test/testp** – 3:2 to 2:3

High definition (--hd)

The **hd** parameter dives into the realms of an early alternative model, generating images with enhanced size but less consistency. While the images are larger and potentially more detailed, there's an inherent variability in the output, making each piece distinct and elevating the depth of abstract and landscape art. It's compatible with model versions 1, 2, and 3.

Figure 4.36 – Using --hd in a prompt using version model 3 versus the same prompt without --hd

Sameseed (--sameseed <number>)

The basic *seed* parameter and the **sameseed** legacy parameter are similar: both use a seed to generate new images and the process of getting the seed number is the same. However, there is a key difference between the two parameters.

The *sameseed* parameter is specific; it generates four very similar images for each seed. The four images generated by the *sameseed* parameter will have the same overall composition, but there will be slight variations in the details. It is only available with model versions 1, 2, 3, `test`, and `testp`.

To get a better idea, let's use exactly the same seed value with `--seed` and with `--sameseed` with the V3 to compare the differences:

/imagine prompt `a growing seed in a beautiful forest --v 3`:

original image   --sameseed 3356599545   --seed 3356599545

Figure 4.37 – Comparing results using the same seed from the original image

Stylize (`--stylize <value between 0 and 1000>` *or* `--s <value between 0 and 1000>`*)*

As previously stated, the *stylize* parameter is compatible across multiple Midjourney versions and models; however, the default and range values vary:

- For model versions 4, 5, and Niji 5, the default value is `100`, and the range is from `0` to `1000`
- For V3, `2500` is the default value, and the range is from `625` to `60000`
- For the `test/testp` versions, `2500` is the default value, and the range is from `1250` to `5000`

Test models (--test, --testp, --creative)

Midjourney, always evolving, periodically unveils **test models** allowing users to explore new artistic horizons. Released over the time when V3 was the default, these three parameters were a game changer.

- `--test`: A general-purpose artistic model emphasizing coherency and creative stylization. Offers stylized and imaginative compositions.
- `--testp`: Specializing in photo-realism, this model also maintains a strong coherence, yielding lifelike images. Prioritizes realism, presenting less stylized but more accurate depictions.
- `--creative`: Can only be used when combined with `--test` and `--testp` for varied and more creative compositions.

> **Note**
> The maximum aspect ratio is 3 : 2 or 2 : 3, influencing the number of initial grid images generated. Two initial grids are generated when the aspect ratio is 1 : 1, or one initial grid if the aspect ratio is not 1 : 1.

a growing seed in a beautiful forest --test

a growing seed in a beautiful forest --test --creative

a growing seed in a beautiful forest --testp

a growing seed in a beautiful forest --testp --creative

Figure 4.38 – Example of grids using --test, --testp, --test --creative, --testp --creative

Upscalers (--uplight, --upbeta)

Although we already covered this in *Chapter 3* under the V3 upscalers, it's important to notice that these are legacy upscaler parameters:

- **Light upscale** (--uplight): --uplight is the smaller upscaler; it produces images with a harmonious and crisp resolution of 1024 x 1024 pixels. It maintains a faithful relationship with the original grid image, resulting in a smoother, less detailed rendition, making it ideal for refining portraits and sleek surfaces in earlier iterations of Midjourney.

- **Beta upscale** (--upbeta): --upbeta summons an alternative beta upscaler. For the preceding model versions (V1–V3), it operates at a resolution of 1024 x 1024 pixels; for subsequent versions, it delivers a sharper 2048 x 2048 pixels output.

***Version** (`--version <number>` **or** `--v <number>`)*

The *version* parameter allows users to use a different version of the Midjourney model on a specific prompt without invoking the **/settings** command. Here is the list of all versions available to date: `--v 1, --v 2, --v 3, --v 4, --v 5, --v 5.1, --v 5.2, --niji 5, --niji 4`.

Exploring the complexities of legacy parameters within Midjourney reveals the enduring creative possibilities provided by settings tailored for older model versions. Every parameter, from aspect ratios and upscale options to model-specific settings, brings forth unique functionalities and limitations, inviting creatives to craft diverse artistic results. While these parameters may not encapsulate the advancements of the most recent models, they stand as enduring elements of Midjourney's ongoing evolution and innovation in the world of AI-generated art.

Summary

This chapter provided a comprehensive guide to understanding the intricate details of parameters in Midjourney, which are essential for creating AI-generated art that aligns closely with your vision.

The chapter clearly explained what parameters are and emphasized their crucial role in influencing the AI's creative output and granting artists greater control and precision.

We embarked on a detailed exploration of the two primary types of parameters: basic and legacy. Basic parameters, the fundamental building blocks of your artistic endeavors, were dissected to reveal their capacity to shape the overall style, composition, and quality of the image. We examined the nuances of *aspect ratio*, *chaos*, *fast*, *image weight*, *quality*, and *style* parameters, each illuminated with practical examples and insights to equip you with a nuanced understanding and hands-on knowledge.

The journey through legacy parameters offered a nostalgic trip back to the roots of Midjourney, acquainting you with the nuances of older model versions and the specificities of using them.

This meticulous exploration and practical insight into parameters is invaluable, ensuring you can navigate the Midjourney tool with increased confidence and expertise, ready to bring your envisioned artworks to life.

As we turn the page to the next chapter, we will embark on the journey of advanced prompting techniques, opening doors to new creative realms. We will learn how to combine images in our prompts in order to obtain an immense variety of creative outputs, explore the fascinating world of multi-prompts, and the possibilities of permutations. This progression promises to build seamlessly on the foundational knowledge from *Chapter 4*, preparing you for a deeper, more enriched exploration of Midjourney's limitless creative possibilities.

Part 3: Advanced Prompting and Visual Creations

This part acts as your command center for creativity within Midjourney, teaching you to craft impactful prompts and utilize Midjourney settings for achieving more precise outcomes. You'll gain the expertise you need to manage intricate prompts and tailor the art-making process, enabling the creation of distinctive and engaging AI art.

This part has the following chapters:

- *Chapter 5, Navigating through Advanced Prompts*
- *Chapter 6, Upgrading Your Prompt for Optimal Results*
- *Chapter 7, Customizing Midjourney – Settings, Preferences, and Unleashing Creative Prompts*

5
Navigating through Advanced Prompts

Now that we have mastered the fundamental concepts of Midjourney, successfully generated images that align with our ideas, and feel comfortable within the tool's domain, it's time to explore how we can better adapt it to our needs and demands. In this chapter, we will learn how to use images in our prompts, decipher the complexity and utility of multi-prompts, and discover how we can refine our creations with the help of a special feature that lets you generate multiple versions of your prompt.

In this chapter, we're going to cover the following topics:

- **Blend mode and image prompting**: In this section, we will learn how to sculpt new creations by skillfully guiding and manipulating the model with image prompts.
- **Multi-prompting**: Here, we will discover and implement the layered richness of multi-prompts for creating visually intricate outputs.
- **Permutations**: Here, readers will learn to use special punctuation to create batches of multiple variations of images from our prompts.

By using these advanced prompting techniques, you can take your Midjourney skills to the next level and create truly unique and innovative images that are more closely aligned with your vision, achieving richer and more detailed visual outputs. This chapter promises to substantially elevate your prowess in crafting unique artwork.

Technical requirements

To make the most of this chapter and follow along with the hands-on activities, you'll need to ensure you've set up everything we covered in *Chapter 2*. Here's a quick recap:

- **Midjourney account**: You'll need an active account on Midjourney. If you haven't registered yet, refer back to *Chapter 2* for a step-by-step guide.

- **Discord installation**: Ensure you have Discord installed on your device, as we'll be working with it alongside Midjourney. Again, the installation guide is available in *Chapter 2*.
- **Payment plans**: At the moment Midjourney has no free or trial plan. A Midjourney paid plan is necessary. Please ensure you've selected the most suitable plan for your needs.
- **Server creation**: Having your own server can significantly enhance and expedite your experience with Midjourney. If you're yet to set up a server in Discord, *Chapter 2* covers the process in detail.

Blend mode and image prompting

Blend mode and **image prompting** are two ways to tell Midjourney what to create using images.

Here are a few examples of what you can do with them:

- Combine images from different genres to create new and unexpected juxtapositions
- Use image prompting to generate variations on existing images or create images in the style of a particular artist or genre
- Use *Blend* mode and image prompting together to create complex and detailed images that would be difficult or impossible to create with text prompts alone

Blend mode and image prompting can help you create amazing images with Midjourney. Let's learn the difference between the two.

Blend mode

Blend mode allows you to combine two to five images into a single new image. It's a user-friendly tool that can be invoked simply by typing **/blend** in the message box:

Figure 5.1 – Type the /blend command in the message box in the same way as you do with the /imagine command

After typing in the command, you can upload up to five images. To do this, simply press **+4 more** and an options box will appear with which you can choose up to three more images to upload and also set the **dimensions**, which can be **Portrait** (2:3), **Square** (1:1) or **Landscape** (3:2). If no dimension is chosen, the generated image will have the default ratio of 1:1.

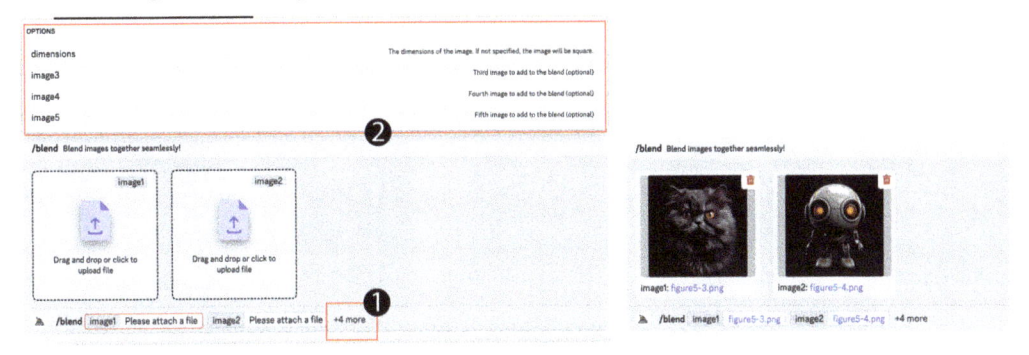

Figure 5.2 – It is optional to use more than two images; note that the more images you use, the more unpredictable the result will be

In our case, we'll choose just two images with the default aspect ratio for now, and hit *enter* to initiate the blending process.

The **/blend** command can be used to create images that are a hybrid of the different concepts and aesthetics of the original images. For example, you could blend an image of a cat with an image of a robot to create a new image of a cyborg cat. The results, while sometimes unpredictable, always offer a unique blend of the input images:

Figure 5.3 – Result of the blending of the two images

Let's test it now with more images and with the portrait dimensions. You can also try making several blends with the same images, simply alternating their order:

Figure 5.4 – Blending five images and using dimensions

Due to the increased use of robot images in our example, the robotic element was significantly amplified in the final outcome:

Figure 5.5 – Result of blending the five images in the portrait aspect ratio

> **Note**
> To get the most out of the blending process, it's advisable to upload images with the same aspect ratio. This ensures a seamless blend and a cohesive final image. Recognizing the growing trend of mobile usage, the **/blend** command is tailored for mobile devices, making it easy for users to drag and drop images or add them from their photo library. That's why it doesn't accept text prompts directly. However, it will accept preferred suffixes (more on this later) if you have defined them.

The *Blend* mode, as we've seen, can be very interesting, but what if we want to add text and get more control over the result? In that case, the most convenient way is to use image prompting. Let's see how this can be done.

Image prompting

The **/imagine** command in Midjourney isn't just for text prompts. By starting your prompt with links to one or more images, you can guide the AI to generate artwork that references these images. This method differs from *Blend* mode because it offers a balance between the aesthetics of the provided images and the direction given by the text prompt.

For instance, blending styles from renowned artists such as Rembrandt or Picasso can produce artwork that captures the essence of *The Weeping Woman* reminiscent of Picasso's style, while retaining elements from the Rembrandt painting *The Night Watch*, on a new water painting canvas:

Figure 5.6 – Using two images and text in a prompt

Let's see how we can get results like the preceding. Start by uploading two previously chosen images. To do this, simply go to the message box, click on + and choose **Upload a File** (in my case, I'm going to choose two images from my computer). Choose your image and click **Open**. Then, click on + again, choose **Upload a File**, repeat the same steps, and press *enter*. Another way is to simply drag the images into the message box and press *enter*:

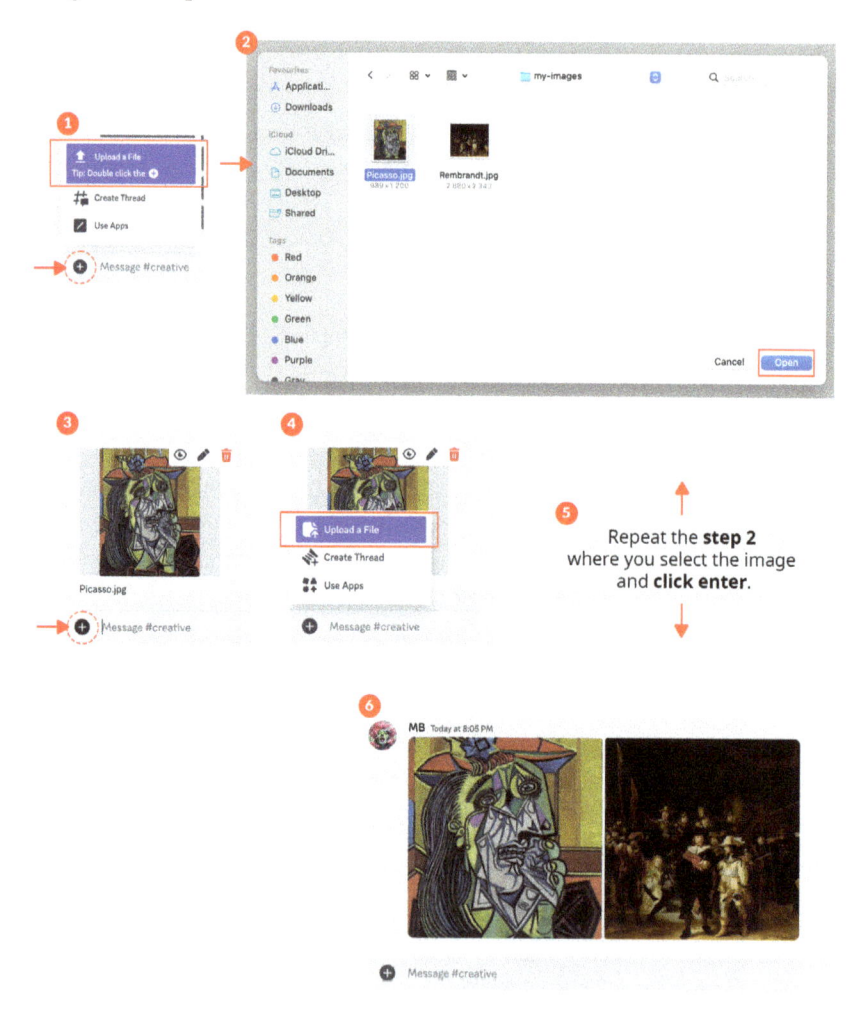

Figure 5.7 – Uploading images to Midjourney

To get links for each image to include them in the prompt, just click on the image you want and a popup with the image and the **Open in Browser** option will appear. Right-click on the image and choose **Copy Image Address**. You can also click on **Open in Browser** and copy the URL that appears:

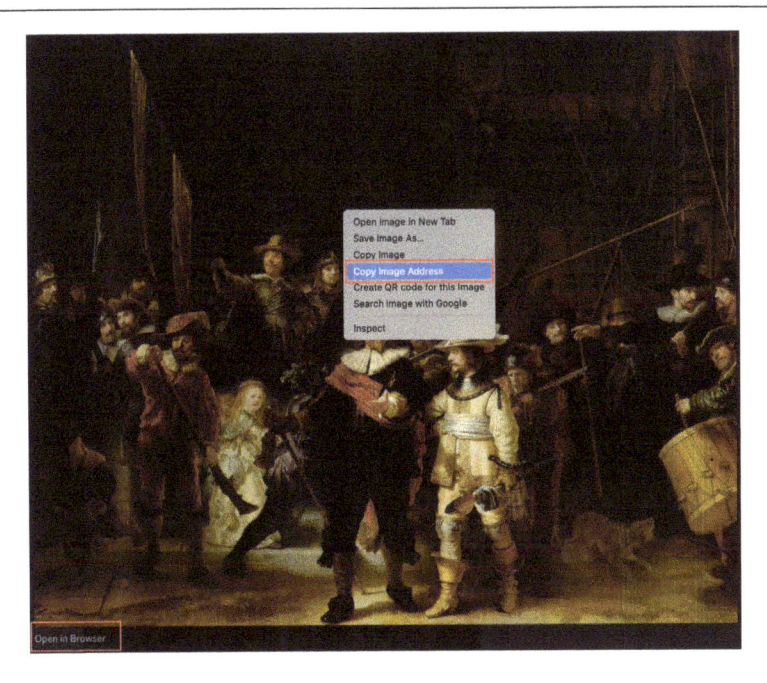

Figure 5.8 – How to find the image URL

Now that I have the URLs of each of my images, I'm going to start composing my prompt. First, I declare the URL of each of my images followed by a description and optionally, one or more parameters:

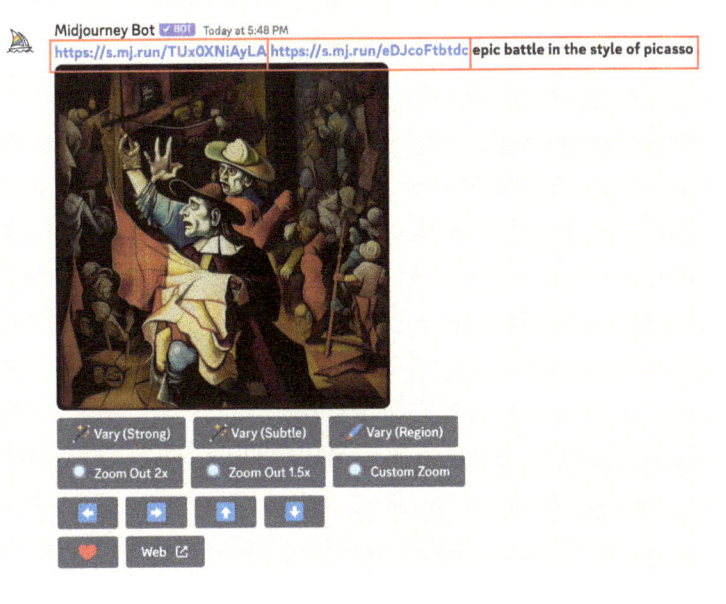

Figure 5.9 – The resulting image generated by using two URLs plus a text prompt

Now let's try something slightly different. We're going to use an image and text, but we're going to introduce a parameter (something we discussed in the previous chapter) – the *image weight* parameter or `--iw`. But first, let's quickly remind ourselves what it's for and how it can help us build our prompts.

One of the standout features of image prompting is the ability to adjust the image weight using the `--iw` parameter. By default, this parameter is set to `1`, ensuring that the influence of the image and text prompt are balanced. However, when you wish to lean more on the visual cues of the reference images, this parameter can be increased, even up to a value of 2. For instance, if I use **/imagine prompt** `<IMAGE URL> text prompt --iw 0.5`, the image prompt will only be half as important as the text prompt. But if I use **/imagine prompt** `<IMAGE URL> text prompt --iw 2`, the image prompt will be twice as important as the text.

What's even more intriguing is Midjourney's capability to assign different weights to various reference images. This means artists can emphasize the aesthetics of one image over another, allowing for a more nuanced final artwork.

Let's use the image of the cat and try to turn it into a robot as we did with the **/blend** command, but this time using image and text:

/imagine prompt `https://s.mj.run/IzyUygkSwEk cute robot with piercing eyes, on a flower pattern background --iw 0.5`

And

/imagine prompt `https://s.mj.run/IzyUygkSwEk cute robot with piercing eyes, on a flower pattern background --iw 2`

The weight and influence of the image in both results is evident, to the point that the style itself is affected: with a lower weight, we get a more illustrative style, while with a higher weight, the photographic aspect of the cat's image dominates in the result:

Figure 5.10 – The same image with the same text in the prompt but with different weights of --iw 0.5 on the left and --iw 2 on the right image

Image prompting in Midjourney is more than just a feature; it's a creative superpower that empowers artists to craft with intent. Image prompting offers a structured approach for using image URLs, descriptive prompts, and parameters to enhance your results, and the possibilities are endless. Let's discover the dynamic world of **multi-prompting**.

Multi-prompting

While the concept of multi-prompting might seem daunting at first, it's a powerful technique in the Midjourney arsenal. Some AI artists have found that as they add more details to their prompts, the results can deviate from their initial expectations, and sometimes this is true. However, it's essential to understand that multi-prompting isn't inherently better or worse than traditional prompting—it's just different. For simple image requests, a standard prompt similar to the ones we've tested and worked with might suffice. But if you are seeking more control and nuance in your creations, multi-prompting offers a level of precision that's hard to match.

Multi-prompting is about dissecting a concept into its elemental components. By employing a double colon (::) as a separator (it's crucial to ensure there's no space between the double colons), you can instruct the Midjourney Bot to interpret each segment of the prompt distinctly. A simple prompt such as `beautiful flowers` might evoke images of a simple composition of flowers. However, segmenting it into `beautiful:: flowers` could lead to a portrait of a beautiful girl surrounded by flowers:

Figure 5.11 – Using the beautiful flowers simple prompt to generate the left image, and the beautiful:: flowers multi-prompt to generate the right image

But multi-prompting isn't just about separation; it's about emphasis. By appending a number immediately after the double colon, we can dictate the prominence of each concept. A prompt like `beautiful::2 flowers` accentuates the beautiful element, leading to images where the themes of beauty and femininity are more pronounced. Both whole numbers and decimals can be used to fine-tune this emphasis depending on the model version (model versions 1, 2, and 3 only accept whole numbers).

Figure 5.12 – Example of an image generated with the weighted multi-prompt beautiful::2 flowers

Sometimes, the goal is to minimize or exclude specific elements. **Negative weights** come into play here. Assigning a negative weight can reduce that element's presence in the final artwork. The `--no` parameter offers a shorthand to this; the `--no` parameter is equivalent to using a negative weight of -.5 (e.g., `your text:: -.5`). For instance, `beautiful::2 flowers --no blue` would produce fewer blue flowers. To test this, I will use the previous image. I click on **Vary (Subtle)** because I want to build on the existing image, and in the pop-up box that appears I add the `--no blues` parameter at the end of the prompt, then hit **Submit**:

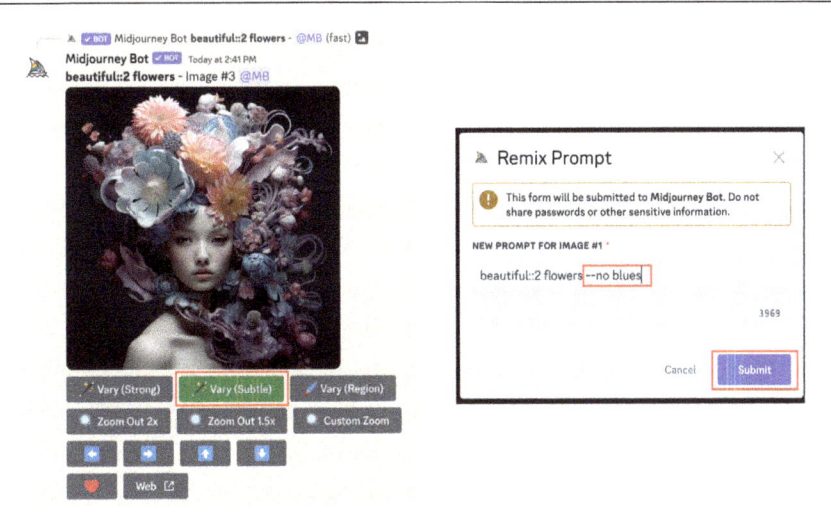

Figure 5.13 – Using the Vary options to edit my multi-prompt and add the --no parameter

> **Weights**
>
> There are two important things to know here: the sum of all the weights must always be positive, and whenever there is an element with no declared weight, it takes on the default weight, which is 1. In other words, `beautiful:: flowers`, `beautiful::1 flowers`, `beautiful:: flowers::1`, and `beautiful::5 flowers::5` are the same in terms of weights.

Now, let's repeat the process to edit the prompt, but instead of using the `--no` parameter, let's use a negative weight. However, we're going to add it at the beginning of the prompt so that it has more impact on the final result: `blues::-1 beautiful::2 flowers`:

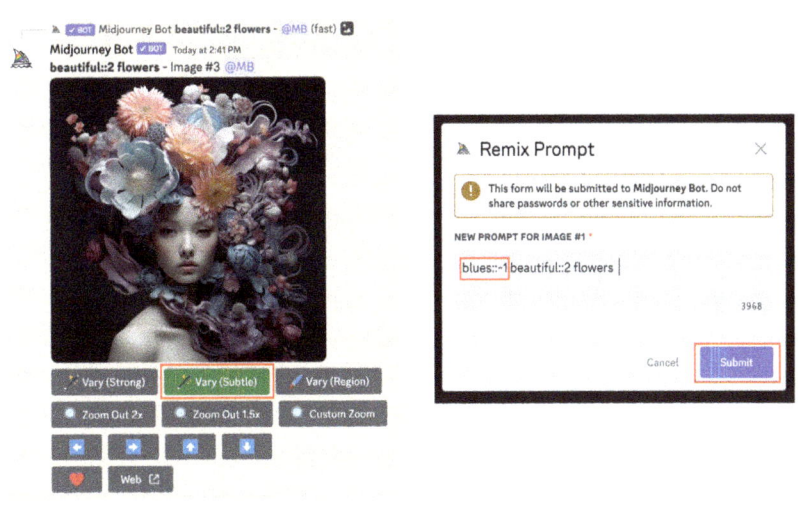

Figure 5.14 – Using the Vary options to edit my multi-prompt and add the negative weight

In terms of consistency in results, I believe using the `--no` parameter is more effective than the negative weight in this case, because as you can see from the following image, the piece on the left has even fewer blue elements. However, in both images, even though the blues are present, they no longer dominate.

Figure 5.15 – Using the --no parameter to generate the left image and the negative weight to generate the right image

The art of multi-prompting lies in the balance. Striking the right amount of detail is key. While being too vague might leave the AI in the dark, overloading it with details can lead to confusion. The best way to master the multi-prompt technique is by following these best practices:

- **Begin with clarity**: Start with a straightforward initial prompt that outlines the primary subject, serving as the image's foundation. Simple prompts such as `beautiful flowers` or `Dragon on a busy NY street` can set the stage. You can then add more information about style, character details, medium, and so on.

- **Consider detailing and balance**: It's a delicate balancing act to ensure the AI understands the intent without overwhelming it with information, so only include details that are essential to your concept. Also keep in mind that the first part of your prompt should have everything you want in your image, and the fewer words you use, the more influence they have.

- **Experiment and use weights**: Multi-prompting is a technique, and you should be prepared for a journey of discovery, refining your prompts based on the outcome and taking more control using weighted prompts. To make it easier, think of weights as percentages of influence.

Multi-prompting | 117

For an engaging culmination of the various topics we've explored so far, I'll combine a multi-prompt with weights and parameters to create a new image prompt.

I'll begin by uploading the robot image (see *Figure 5.2*) previously used in the *Blend* mode example, and pair it with one of the images I generated more recently (see *Figure 5.15*). Each image will be assigned a distinct weight.

The image prompt is as follows: **/imagine prompt** `<IMAGE URL FROM MY ROBOT IMAGE>::1.25 <IMAGE URL FROM MY BEAUTIFUL FLOWERS>::2`:

Figure 5.16 – Result of blending the two images using weights

The following figure is the result generated using the same images and weights as previously, but with the addition of the text prompt `low poly digital art` and an *aspect ratio* parameter of `--ar 3:2`. We can make an interesting observation here: even though the robot image has a weight of `1.25`, its influence on the final generated image is minimal. The text prompt, which specifies a new style, plays a more dominant role. Typically, a higher image weight emphasizes its visual features in the output, while a lower weight accentuates the text prompt. Yet, the image of the girl with flowers with a weight of 2 is more evident in the final output. This is because the weight is relative, making the girl's image naturally stand out more than that of the robot.

Figure 5.17 – Image generated by using <IMAGE URL FROM MY ROBOT IMAGE>::1.25 <IMAGE URL FROM MY BEAUTIFUL FLOWERS>::2 low poly digital art --ar 3:2

Overall, we can say that multi-prompting is a powerful technique for generating specific and nuanced images with Midjourney. It can be used for emphasizing a character or a style, filtering unwanted elements, blending multiple artistic concepts into an image, and much more. By understanding how to use weights and separators, you can create prompts that instruct the AI to generate images that are tailored to your specific vision.

Now that we have explored the advantages of multi-prompting, we can further refine our creative output by diving into permutations, which is a technique that can maximize our creation of images and let us test a variety of options.

Permutations

Permutations in Midjourney enable us to generate multiple image variations from a single prompt. By using special punctuation, specifically curly brackets { }, we can instruct the AI to produce different combinations that we specify. This feature is akin to coding, where specific markers tell the system how to interpret and execute commands.

For instance, **/imagine prompt** A `{woman, man} in a {forest, city} during {day, night}` will yield multiple image variations, such as a woman in a forest during the day, a man in the city at night, and so forth.

The true strength of permutations lies in its flexibility. Let's see some of the ways you can maximize the potential of permutations.

Permutations and parameters

Permutations allow for the adjustment of parameters and weights within a prompt. This means you can experiment with different aspect ratios, versions, or even chaos values, all within a single prompt. For weight variations, make sure you don't leave space between the two dots and the curly brackets (e.g., **/imagine prompt** `text text text::{1, 2.}`).

Here's an example with different aspect ratios and versions: **/imagine prompt** `rice field in China, harmonious ambience, little mist in the air, beautiful atmosphere, realism --ar {1:1, 3:2, 2:3, 1:2, 16:9} --v {5, 5.2}`.

Note that in the case of permutations, after typing your prompt, a message will appear asking whether you are sure you want to create images of all the prompt permutations. The preceding prompt will produce 10 results and I can choose from the following options:

- **Yes** (i.e., go ahead and generate the 10 individual permutations)
- **No** (i.e., cancel the generation)
- **Show Prompts**, which lets me see the list of individual prompts that will be generated
- **Edit Template**, which allows me, similar to *Remix* mode, to edit my prompt via a pop-up box

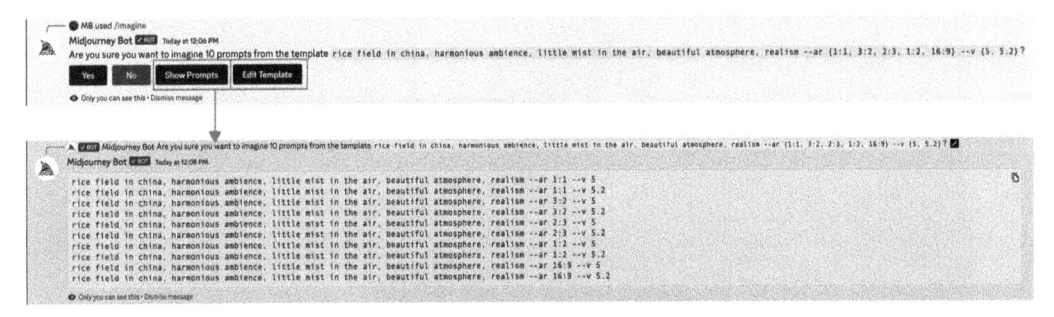

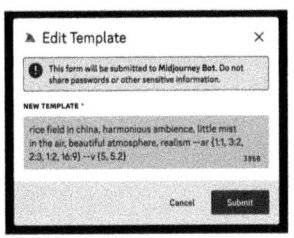

Figure 5.18 – The Midjourney Bot will show a confirmation message before the permutation prompts begin processing

In my case, I hit **Yes** and generate all the permutations:

- ```
  rice field in China, harmonious ambience, little mist in the
  air, beautiful atmosphere, realism --ar 1:1 –v 5:
  ```

Figure 5.19 – Grid of four images generated by using --ar 1:1 --v 5

- ```
  rice field in China, harmonious ambience, little mist in the
  air, beautiful atmosphere, realism --ar 1:1 --v 5.2:
  ```

Figure 5.20 – Grid of four images generated by using --ar 1:1 --v 5.2

- ```
  rice field in China, harmonious ambience, little mist in the
  air, beautiful atmosphere, realism --ar 3:2 --v 5:
  ```

Figure 5.21 – Grid of four images generated by using --ar 3:2 --v 5
  ```

122　Navigating through Advanced Prompts

- ```
  rice field in China, harmonious ambience, little mist in the
  air, beautiful atmosphere, realism --ar 3:2 --v 5.2:
  ```

Figure 5.22 – Grid of four images generated by using --ar 3:2 --v 5.2

- ```
  rice field in China, harmonious ambience, little mist in the
  air, beautiful atmosphere, realism --ar 2:3 --v 5:
  ```

Figure 5.23 – Grid of four images generated by using --ar 2:3 --v 5

- ```
  rice field in China, harmonious ambience, little mist in the
  air, beautiful atmosphere, realism --ar 2:3 --v 5.2:
  ```

Figure 5.24 – Grid of four images generated by using --ar 2:3 --v 5.2
  ```

- `rice field in China, harmonious ambience, little mist in the air, beautiful atmosphere, realism --ar 1:2 --v 5:`

Figure 5.25 – Grid of four images generated by using --ar 1:2 --v 5

- `rice field in China, harmonious ambience, little mist in the air, beautiful atmosphere, realism --ar 1:2 --v 5.2:`

Figure 5.26 – Grid of four images generated by using --ar 1:2 --v 5.2

- ```
  rice field in China, harmonious ambience, little mist in the
  air, beautiful atmosphere, realism --ar 16:9 --v 5:
  ```

Figure 5.27 – Grid of four images generated by using --ar 16:9 --v 5
  ```

- ```
  rice field in China, harmonious ambience, little mist in the
  air, beautiful atmosphere, realism --ar 16:9 --v 5.2:
  ```

Figure 5.28 – Grid of four images generated by using --ar 16:9 --v 5.2

Styles and aesthetics

You can blend different artistic styles and aesthetics seamlessly. Continuing with the previous example, I now want to experiment with camera angles, so I want to ask for permutations of results if the image were a photo taken from the air. So, my prompt will be **/imagine prompt** `{aerial view, drone-footage, elevated shot} rice field in China, harmonious ambience, little mist in the air, beautiful atmosphere, realism --ar 3:2`.

Once again, I get a number of variations and can quickly choose the one I like best and that best fits my requirements:

Figure 5.29 – Grid of four images generated using aerial view rice field in China, harmonious ambience, little mist in the air, beautiful atmosphere, realism --ar 3:2
  ```

Figure 5.30 – Grid of four images generated using drone-footage rice field in China, harmonious ambience, little mist in the air, beautiful atmosphere, realism --ar 3:2

Figure 5.31 – Grid of four images generated using elevated shot rice field in China, harmonious ambience, little mist in the air, beautiful atmosphere, realism --ar 3:2

Nested permutations

Nested permutations offer even more intricate combinations. By nesting curly brackets within others, we can generate a plethora of image variations. Here is the example with the prompt, **/imagine prompt** `elevated shot of a {rice field in {India, Africa}, harmonious ambience, {little mist, rain} in the air}, beautiful atmosphere, realism --ar 3:2`. This will generate five different variations:

Figure 5.32 – Grid of four images generated using elevated shot of a rice field in Africa, beautiful atmosphere, realism --ar 3:2

Figure 5.33 – Grid of four images generated using elevated shot of a harmonious ambience, beautiful atmosphere, realism --ar 3:2

Figure 5.34 – Grid of four images generated using elevated shot of a rice field in India, beautiful atmosphere, realism --ar 3:2

Figure 5.35 – Grid of four images generated using elevated shot of a little mist in the air, beautiful atmosphere, realism --ar 3:2

Figure 5.36 – Grid of four images generated using elevated shot of a rain in the air, beautiful atmosphere, realism --ar 3:2

Weights and image references

Permutations also allow us to test different weight variations within a prompt. To create permutations of weight values in a prompt to get more control over specific elements, use a format such as **/imagine prompt** `text prompt::{1, 2.}`, and make sure you don't leave a space between the two dots and the curly brackets. We can also use reference images in permutations, as follows: `{IMAGE URL, IMAGE URL} {IMAGE URL, IMAGE URL} your text prompt`.

> **Note**
>
> Permutation prompts are only available in *Fast* mode and each variation of the permutation counts as an individual job. With a single permutation prompt, Basic subscribers can create a maximum of 4 jobs; Standard subscribers, a maximum of 10 jobs; and *Pro* and *Mega* subscribers, a maximum of 40 jobs. Be mindful of GPU usage: as already mentioned, each job in *Fast* mode consumes GPU minutes. For more information, check your subscription.

As is clear by now, permutations serve to speed up our creative process. You can get variations of practically everything in your prompt very quickly. Whether you're adjusting parameters, experimenting with styles, or blending different aesthetics, permutations offer a world of possibilities almost at the speed of light.

Summary

In this chapter, we showcased the innovative features of Midjourney, starting with *Blend* mode and image prompting. These tools empower artists to seamlessly combine images, allowing for unique and creative results. We then transitioned into multi-prompting, a technique that offers a more controlled approach to blending using multiple text prompts. This method provides artists with the flexibility to dictate the style, genre, and medium of their creations. Lastly, we explored permutations, a groundbreaking feature that unlocks a multitude of creative options from a single prompt, offering artists the chance to experiment with endless variations.

In the next chapter, we'll dive into fine-tuning prompts for optimal results. We'll explore styles, introduce new features such as *describe*, and uncover techniques to further refine and perfect your artistic endeavors in Midjourney.

6
Upgrading Your Prompt for Optimal Results

Crafting the perfect prompt in Midjourney is much like molding clay into a beautiful sculpture – it demands precision, understanding, and the right techniques. As we progress in our journey through AI art creation, it becomes essential to use the full potential of the tool for more detailed and controlled outputs. This chapter is designed to take your prompting skills a notch higher by introducing you to advanced techniques and features.

To begin with, we'll embark on a journey through the *Help me describe* section, where we will discover a gem – the **describe** feature – which offers a richer artistic vocabulary, thus broadening the horizons of our creative expression.

We then transition to the myriad of styles available. The power of style selection can't be understated – it can be the difference between a good image and an extraordinary one. But it doesn't stop there – constructing an effective prompt goes beyond just knowing styles and descriptions. You will understand the essence of choosing the right words as well as grasp how tools such as **ChatGPT** and the **shorten** feature can be crucial in crafting these perfect prompts.

In this chapter, we're going to cover the following topics:

- **Help me describe**: In this section, you will learn how to use richer and more nuanced artistic vocabulary for enriching prompts.
- **A world of styles**: Here, you will discover the various styles available and choose the one that best suits your desired outcome.
- **The right words**: This section will show how you can craft the ideal prompt using essential words, with assistance from ChatGPT and the *shorten* feature

By the end of this chapter, you will master the art of prompting in Midjourney. You'll learn how to spice up your prompt with the *describe* feature, helping you to expand your artistic vocabulary and bring more nuance to your prompts. We'll explore the diverse world of styles, showing you how to pick just the right one for the vibe you're going for. And it doesn't stop there. With insights into choosing the right words, tools like ChatGPT, and the *shorten* feature, you'll learn how to fine-tune your prompts for obtaining the best results. By the end, you'll know how to get exactly what you want from Midjourney, making your art stand out even more.

Technical requirements

To make the most of this chapter and follow along with the hands-on activities, you'll need to ensure you've set up everything we covered in *Chapter 2*. Here's a quick recap:

- **Midjourney account**: You'll need an active account on Midjourney. If you haven't registered yet, refer back to *Chapter 2* for a step-by-step guide.
- **Discord installation**: Ensure you have Discord installed on your device, as we'll be working with it alongside Midjourney. Again, the installation guide is available in *Chapter 2*.
- **Payment plans**: At the moment, Midjourney has no free or trial plan. A Midjourney paid plan is necessary. Please ensure you've selected the most suitable plan for your needs.
- **Server creation**: Having your own server can significantly enhance and expedite your experience with Midjourney. If you're yet to set up a server in Discord, *Chapter 2* covers the process in detail.

Help me describe

Have you ever wondered what it would be like to have an AI assistant who could help you to create stunning images? With the *describe* command, you can do just that!

I like to give the following examples, so one can understand the benefits of using the **/describe** flag. Imagine you come across a mesmerizing image on the web and wish to craft something similar on Midjourney, yet you find yourself at a loss for words. Or perhaps you've been curious about the inner workings of the Midjourney bot, pondering over the logic behind its image creations. Maybe there are times when your creative well runs dry, and you're in dire need of fresh inspiration. If any of these situations sound familiar, then the *describe* command is your new best friend.

The **/describe** flag allows you to upload an image and receive four distinct prompts based on the visual information that Midjourney gleans from it. These prompts can then be used as inspiration for new images, or to help you to develop your own artistic style. However, it's essential to note that these generated prompts are of an inspirational nature, suggestive at best, rather than precise blueprints. They don't serve to replicate the uploaded image but instead provide a unique perspective on it. A bonus? The *describe* command even gives you the aspect ratio of the uploaded image.

Here are some of the benefits of using the *describe* command:

- Discover new artistic styles
- Learn how to create more sophisticated prompts

- Learn how Midjourney prefers the use of words
- Generate fresh inspiration when you're feeling stuck

To try out the *describe* command, simply type /**describe** in the chat box:

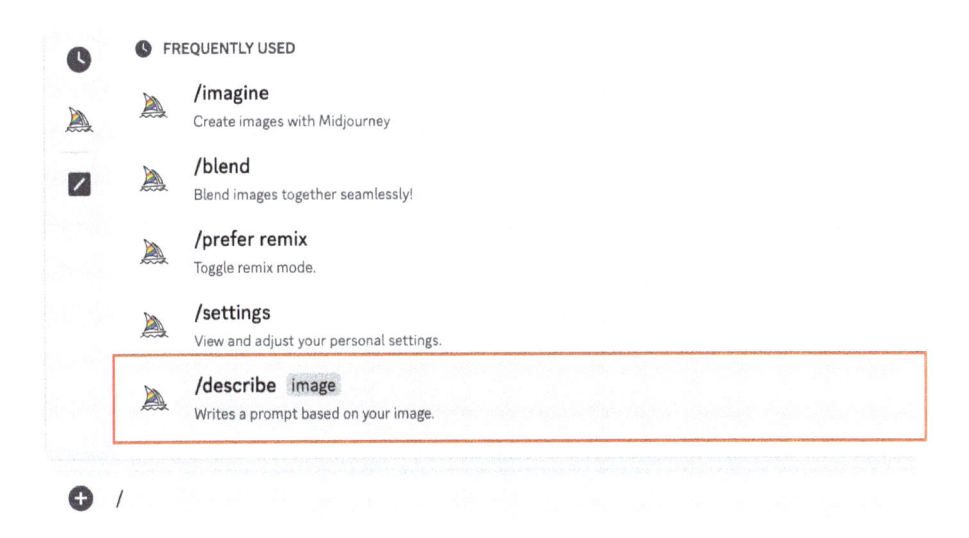

Figure 6.1 – Type the /describe command in the message box in the same way as you do with the /imagine command

Upload the image you want to describe and hit *enter*:

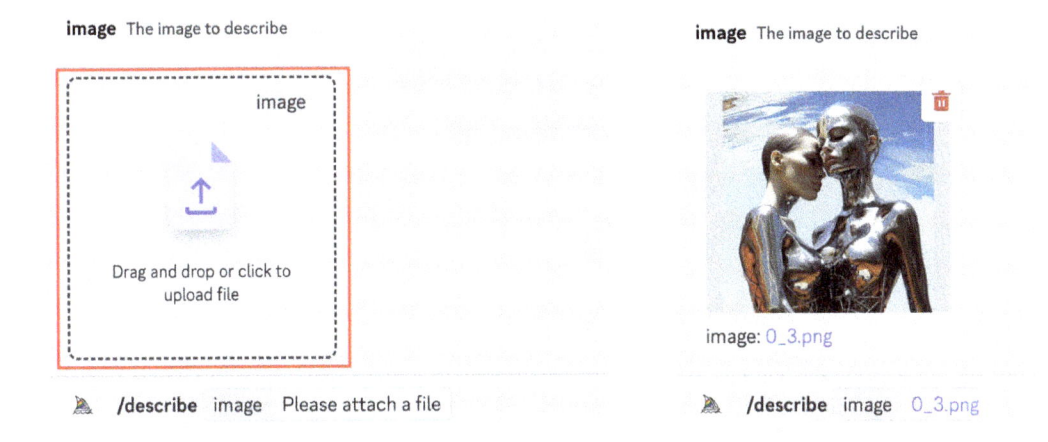

Figure 6.2 – Upload the image you want to describe

Midjourney will then generate four prompts for you. Now what we can do is click on each of the numbers that appear below the image and choose the ones we want to generate, or we can choose **Imagine all**, which will start generating all the images. If you're not happy with the prompts or curious to find more examples, you can ask Midjourney to re-generate new prompt proposals; to do this, just do the normal re-roll by pressing the last button (circular arrows on the far right):

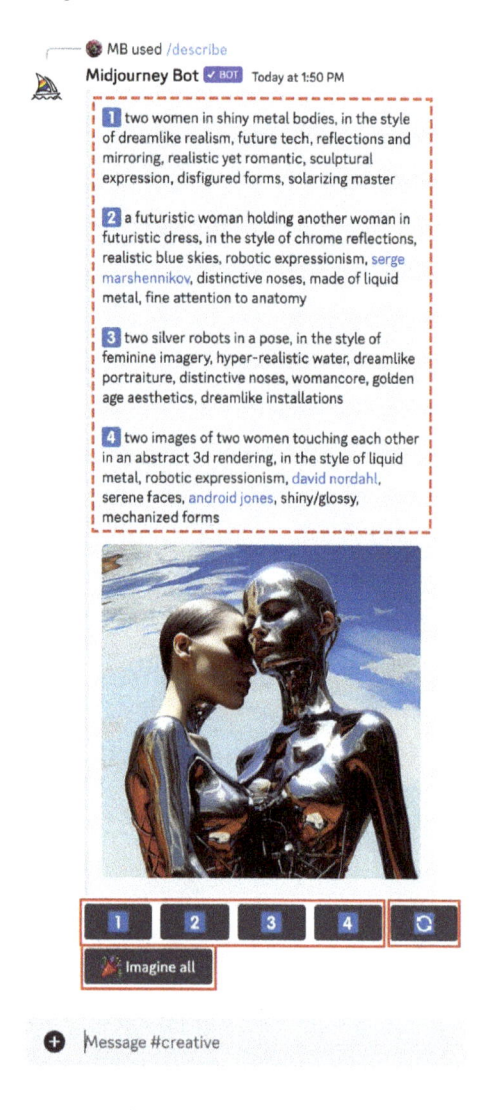

Figure 6.3 – Examples of Midjourney prompts generated with the /describe flag

In the following figure, you can see the result of the four-image grids generated by each of the prompts:

Figure 6.4 – Grids created with each prompt

It's interesting to note that the image I uploaded was one I had previously generated using Midjourney.

In other words, what I did here was *reverse engineering* in the sense that I wanted to understand how Midjourney would interpret my image, and if it had to create it, which prompt it would use.

For the sake of clarity, the prompt that generated the initial image is none other than a re-roll of an image generated using **/imagine prompt** `woman and an eternal shiny chrome robot in the future`:

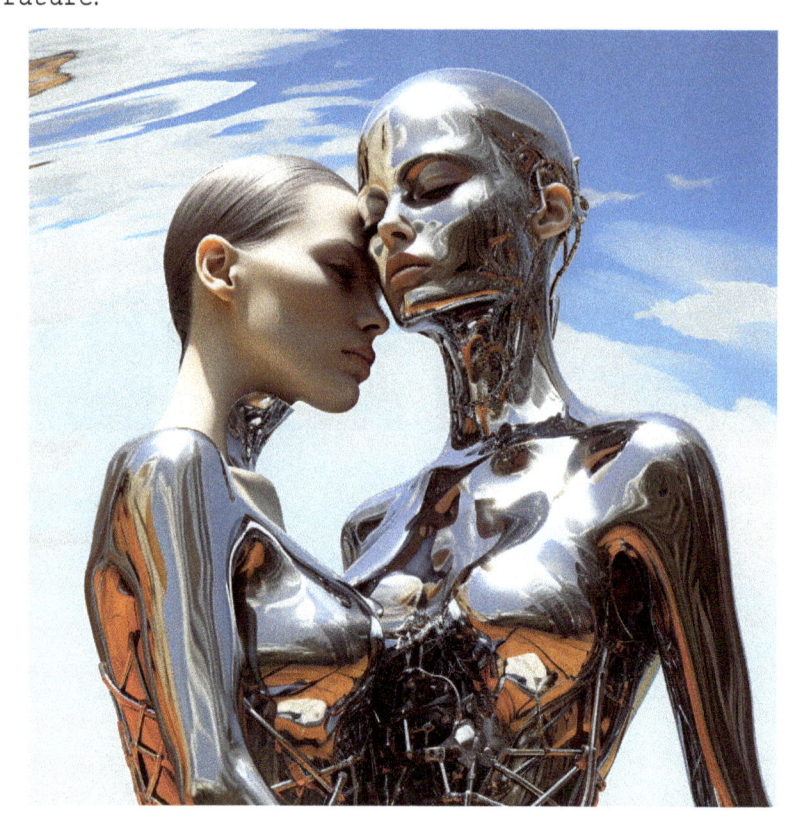

Figure 6.5 – Original image used for this describe exercise

Tip

Want results that closely resemble your uploaded image (see *Figure 6.5*)? Combine the *describe* command with an image prompt, a technique we touched upon in *Chapter 5*. Here's how:

1) Use the *describe* command to get a list of descriptions, as we saw before.

2) Copy your preferred description.

3) Upload the same image you used for the *describe* command to Midjourney and obtain its link URL.

4) Craft a new prompt by combining the image link and your chosen description. Note that you can also add your own words or styles to adjust the prompt even better.

Figure 6.6 – Image prompt with the result of the describe command to generate an image closer to the original one

For instance, I opted for the second description. The resulting grid, as you'll observe, aligns more with the original image compared to solely using the describe command:

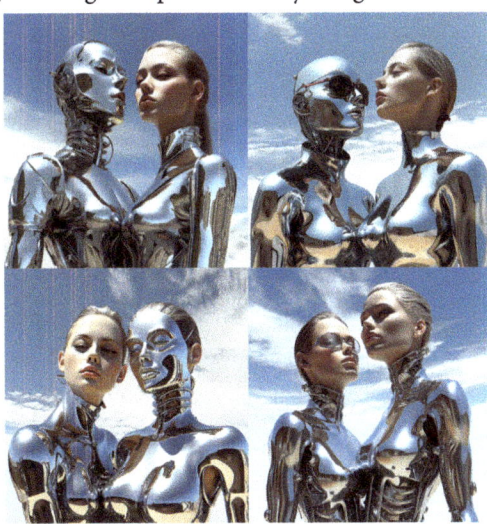

Figure 6.7 – Image prompt and describe result

In the end, when examining the prompt proposals generated by Midjourney, we can observe several key patterns. First, Midjourney prioritizes certain words, always placing them first in the prompt (e.g., *two*, *woman*, *chrome*, or *metal*). Second, it introduces new terms such as *solarizing master* and *robotic expressionism*, and even artist names (e.g., *Serge Marshennikov*, *David Nordahl*, and *Android Jones*), and defines styles (e.g., *in the style of dreamlike realism, in the style of liquid*, and so on). This suggests that Midjourney is capable of understanding and interpreting the nuances of human language and that it uses this knowledge to generate prompts that are both creative and informative.

By incorporating these new ideas into our prompts, we can improve the results of our image creations and expand the range of possibilities. In the next section, we will explore how to add styles to our prompts to achieve even greater results.

A world of styles

Adding styles to Midjourney prompts is a great way to improve the results of your image creations and expand the range of possibilities. When you specify a style, Midjourney will try to generate an image that is consistent with that style. This can be helpful if you are trying to create a specific type of image, such as a landscape in the style of impressionism or a portrait in the style of Renaissance art.

There are two main ways to use styles in Midjourney prompts:

- **Using the "in the style of" phrase**: This is the most straightforward way to specify a style. Simply type `in the style of` followed by the style you want, for example, *in the style of impressionism* or *in the style of Renaissance portraits*.

- **Using keywords**: Midjourney also understands keywords. For example, if you want to create an image in the style of expressionism, you could use keywords such as *bold colors*, *distorted forms*, or *emotional intensity*.

You can also combine the two methods to create even more specific prompts. For example, you could say `a portrait of a woman in the style of Renaissance art, with bold colors and a dreamy atmosphere`.

Here are some examples of how using styles can produce different results:

- **Basic prompt**:

    /imagine prompt `a portrait of a woman`

    Output: Midjourney will choose a style based on its own knowledge and understanding of the prompt:

Figure 6.8 – Image generated using just the words "a portrait of a woman"

- **Prompt with "in the style of"**:

    /imagine prompt `a portrait of a woman in the style of Renaissance art`

    Output: This image will likely have realistic proportions, idealized features, and a focus on facial features and beauty or clothes ornaments:

    Figure 6.9 – Image generated using a specific style

- **Prompt with keywords**:

    /imagine prompt `a portrait of a woman with bold colors and a dreamy atmosphere`

    Output: This image may have saturated colors and a sense of fantasy or otherworldliness:

    Figure 6.10 – Image generated using keywords

- **Prompt with both "in the style of" and keywords**

    /imagine prompt `a portrait of a woman in the style of Renaissance art, with bold colors and a dreamy atmosphere`

    Output: In this case, the image may be closer to the Renaissance art style than the previous prompt, but with more detail and a more ethereal atmosphere:

    Figure 6.11 – Image generated using both "in the style of" and keywords

We can use styles and keywords to create and define a wide range of images, encompassing various artistic styles from traditional fine art to digital art, and across different mediums, including canvas, digital platforms, or photography. It's important to know that in addition to the name of the styles and the keywords, any other word or expression that reinforces the style can further direct the result toward what we want, but we can also use simple prompts such as **/imagine prompt** `<subject>, in the style of <Avant-garde, Art Deco, Art Nouveau, Baroque, Bauhaus, Geometric, Gothic, Minimalism, Ukiyo-E>`. We'll see some examples of specific art styles and keywords.

Art styles

- **Anime and manga**: Anime is characterized by vibrant colors, exaggerated facial expressions, and dynamic movement, while a manga style generates images that are more detailed and intricate, akin to traditional Japanese comic art:

Figure 6.12 – /imagine prompt - anime shot of women dressed in floral shirts posing for a magazine, in the style of anime, dark orange and light bronze, wrapped, intense close-ups, luxurious fabrics

- **Cartoon**: Simplistic, childish visuals with bright hues and minimalistic details:

Figure 6.13 – /imagine prompt - cartoon of women dressed in floral shirts posing for a magazine, in the style of cartoon, dark orange and light bronze, wrapped, intense close-ups, luxurious fabrics

- **Comic book**: These images are similar to cartoons, but they have a more dynamic and stylized look with bold lines:

Figure 6.14 – /imagine prompt - comic strips of women dressed in floral shirts posing for a magazine, in comic book style with bold lines, dark orange and light bronze, wrapped, intense close-ups, luxurious fabrics

- **Ink drawing**: These images are created using ink and paper, and they often have a bold, fluid, and expressive style:

Figure 6.15 – /imagine prompt - BW ink drawing of women dressed in floral shirts posing for a magazine, in ink drawing style, fine lines black and orange, wrapped, intense close-ups, luxurious fabrics, pen and ink drawing technique

- **Oil painting**: Rich and textured visuals crafted with oil paints, reminiscent of Rembrandt's masterpieces:

Figure 6.16 – /imagine prompt - oil painting of women dressed in floral shirts posing for a magazine, in oil painting style, dark orange and light bronze, wrapped, intense close-ups, luxurious fabrics

- **Pastel drawing**: These images are created using pastels, and they often have a soft and delicate look:

Figure 6.17 – /imagine prompt - Pastel drawing of women dressed in floral shirts posing for a magazine, in pastel drawing style, dark orange and light bronze, wrapped, intense close-ups, luxurious fabrics

- **Pencil drawing**: These images are created using pencils, and they often have a realistic and detailed look:

Figure 6.18 – /imagine prompt - Pencil drawing of women dressed in floral shirts posing for a magazine, in pencil drawing style, dark orange and light bronze, wrapped, intense close-ups, luxurious fabrics

- **Watercolor painting**: Known for their transparency and fluidity, these images are created using watercolor paints, and they often have a soft and ethereal look:

Figure 6.19 – /imagine prompt - Watercolor painting of women dressed in floral shirts posing for a magazine, in watercolor painting style, dark orange and light bronze, wrapped, intense close-ups, luxurious fabrics

- **Sci-Fi**: These images showcase futuristic settings with advanced tech and alien life. We can adjust the prompt, for example, by inserting movie names, which can serve as inspiration and guide the final result. It also works well with words such as *retro* and futuristic terms:

A world of styles 151

Figure 6.20 – /imagine prompt - Futuristic shot of women dressed in floral shirts posing for a magazine, in the style of futuristic Sci-Fi, dark orange and light bronze, wrapped, intense close-ups, luxurious fabrics

- **Pixel art**: Digital art reminiscent of early video game graphics. We can use terms such as *8-bit pixel art* or console, game, and character names. Interestingly, I've noticed that the pixel art works better on the V4 model, perhaps because V5 is more realistic and detailed:

Figure 6.21 – /imagine prompt - 8-bit pixel art of women dressed in floral shirts posing for a magazine, in the style of Legend of Zelda: A Link to the Past (SNES, Game Boy Color) 1991, dark orange and light bronze, wrapped, intense close-ups, luxurious fabrics --v 4

- **Vector art**: This style uses mathematical formulas to create images, resulting in sharp and scalable graphics. It's important to add keywords that are common in the design world to reinforce the style. Some examples are *silhouette*, *vector*, *flat*, *graphic vector*, *logo*, and *white background*:

Figure 6.22 – /imagine prompt - vector logo of women dressed in floral shirts posing for a magazine, in the style of minimalistic vector art, dark orange and light bronze, wrapped, intense close-ups, luxurious fabrics, isolated on a white background

- **Low poly**: These 3D images are characterized by simple, geometric shapes:

Figure 6.23 – /imagine prompt - Low poly shot of women dressed in floral shirts posing for a magazine, in the style of low poly, dark orange and light bronze, wrapped, intense close-ups, luxurious fabrics

- **Glitch art**: This style intentionally introduces digital errors into images, creating a distorted and otherworldly effect:

Figure 6.24 – /imagine prompt - Glitchcore effect, breakcore, women dressed in floral shirts posing for a magazine, in the style of Glitch art, dark orange and light bronze, wrapped, intense close-ups, luxurious fabrics

- **Cyberpunk**: These images depict dystopian settings with advanced technology:

Figure 6.25 – /imagine prompt - Cyberpunk women dressed in floral shirts posing for a magazine, in the style of cyberpunk, neon luminescent colors, wrapped, intense close-ups, luxurious fabrics

- **Pop art**: Featuring imagery from popular culture and mass media:

Figure 6.26 – /imagine prompt - Pop art painting of women dressed in floral shirts posing for a magazine, in the style of Roy Lichtenstein pop art, dark orange and light bronze, wrapped, intense close-ups, luxurious fabrics

A world of styles 157

Mood styles

In addition to the specific art styles mentioned in the previous subsection, Midjourney can also generate images with more general stylistic qualities, such as the following:

- **Moody**: These images have a dark, heavy, and atmospheric feel:

Figure 6.27 – /imagine prompt - moody comic book style illustration of a panda

- **Evocative**: Suggestive and thought-provoking visuals:

Figure 6.28 – /imagine prompt - evocative comic book style illustration of a panda

- **Magical realism**: These images combine realistic elements with fantastical elements:

Figure 6.29 – /imagine prompt - ethereal comic book style illustration of a panda in the style of magical realism

Photographic styles

Since photography is a unique medium for capturing the essence of reality, we can also create images in a wide range of photographic styles, such as food photography, or editorial, studio, or commercial photography, each of which offers a different perspective. Here are some of these distinct photography styles with a few sample images:

- **Portrait photography**: This style emphasizes capturing the essence and personality of an individual or group. It's about more than just a face; it's about telling a story through a person's expressions and demeanor:

Figure 6.30 – /imagine prompt - portrait of a middle-aged couple smiling, vintage photo

- **Landscape photography**: This style captures the beauty of natural and man-made environments. From majestic mountains to bustling cityscapes, it's about showcasing the world in all its grandeur (see *Figure 6.47*).
- **Still life photography**: A genre that focuses on inanimate subjects, often arranged in a thoughtful composition. It's a way to highlight the beauty and significance of everyday objects (see *Figure 6.40*).

- **Knolling photography**: The process of organizing and arranging objects in parallel or at 90-degree angles:

Figure 6.31 – /imagine prompt - knolling photography of gourmet spices and ingredients

- **Nature photography**: This style celebrates the beauty of the natural world, from flora and fauna to the changing seasons and weather patterns (see *Figure 6.47*).

- **Wildlife photography**: A challenging yet rewarding style that captures animals in their natural habitats, showcasing their behaviors, interactions, and the environments they inhabit (see *Figure 6.41*).

- **Cityscape photography**: Highlighting the architectural wonders and the hustle and bustle of urban life, this style captures cities in all their glory (see *Figure 6.38*).

- **Astrophotography**: A mesmerizing style that focuses on celestial events and objects, from starry night skies to awe-inspiring planetary events (see *Figure 6.48*).

- **Macro photography**: This style magnifies the minute, revealing the intricate details of small subjects that might be overlooked by the naked eye:

Figure 6.32 – /imagine prompt - Macro photography of a moth in flight, water droplets flying through the air, highly detailed, high-speed photography, cinematic

- **Aerial photography**: Offering a bird's-eye view, this style captures landscapes and subjects from above, often using drones, providing a fresh perspective on familiar scenes (see *Figure 6.47*).

- **Underwater photography**: Delving into the depths, this style showcases the wonders of marine life and underwater landscapes, revealing a world that's often out of sight:

Figure 6.33 – /imagine prompt - Underwater photography of exotic fish, in the style of flowerpunk, volumetric lighting

- **Street photography**: Candid and spontaneous, this style captures the pulse of urban life, from fleeting moments to the myriad stories unfolding on the streets (see *Figure 6.39*).
- **High-speed photography**: Captures the essence of motion, revealing the invisible world at speeds imperceptible to the human eye:

Figure 6.34 – /imagine prompt - high-speed photography of city traffic lights --ar 4:2

- **Double exposure photography**: A creative technique that blends two or more images, resulting in surreal and dreamlike compositions that challenge our perception of reality:

Figure 6.35 – /imagine prompt - forest, city, double exposure photography

- **Photogram**: A creative technique where a photographic image is created without a camera, by placing objects on light-sensitive paper and exposing it to light:

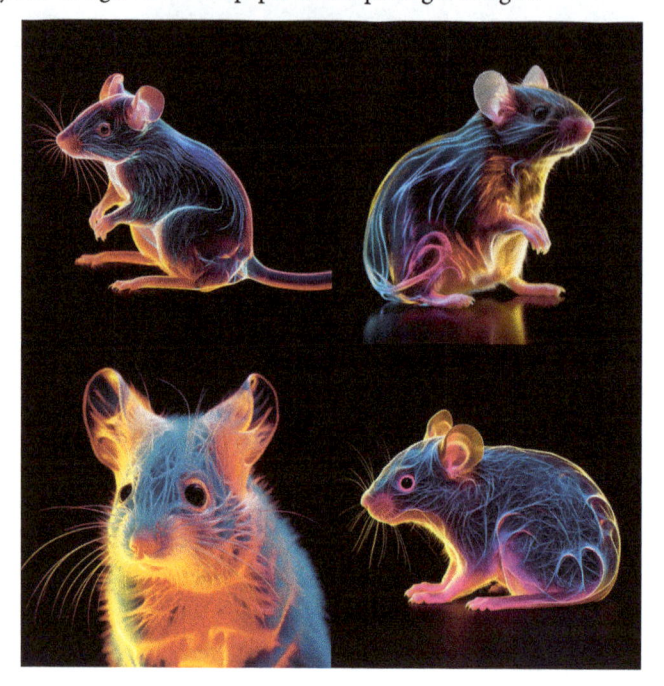

Figure 6.36 – /imagine prompt - rgb photogram of a mouse

> **Art style guidance tips**
>
> Achieving the desired art style with Midjourney can depend on various factors. While a single word can alter the artwork's direction, achieving precise results often requires more finesse. Here are some strategies:
>
> - **Enhance your prompt**: Instead of just adding a style descriptor such as *surreal*, consider including related terms such as *dreamscape* or *floating* to guide Midjourney more accurately.
>
> - **Experiment with aspect ratios**: Some styles, especially panoramic landscapes, might fit wide frame shapes better. For a broad view, you might use `--ar 16:9`.
>
> - **Incorporate artist references**: I personally don't use these, but you can get some interesting results by mentioning artists, such as *Picasso* for a Cubist style. Combining names, such as *Dali* and *Magritte*, can produce unique blends.
>
> - **Core aesthetics**: These are style descriptors that can be used to influence Midjourney's output. For example, try appending *core* to any aesthetic description and see what happens. You can also combine core prompt aesthetics with artist references to create even more unique results. Examples of some core aesthetics are *flowercore, bubblecore, punkcore, ghostcore, dragoncore,* and *naturecore*.

Light

In addition to photographic styles, there is also a huge variety of lighting and camera styles that can positively affect your generated image. Here is a list of the most common types of lighting:

- **Volumetric lighting**: This type of lighting creates the illusion of depth and realism by simulating the way light interacts with objects in space. Here are some examples of keywords: *volumetric light*, *god rays*, and *scenic lighting* (this type of lighting is used for scenic shots and can be uplifted if combined with atmospheric conditions: *natural lighting*, *colorful photo*, *bright sunlight*, and *rays of light*):

Figure 6.37 – /imagine prompt - a breakfast scene bathed in gentle morning sunlight Include delicious chocolate muffins with a golden-brown top and molten chocolate chips, placed near a bowl of crispy breakfast cereal, volumetric light with a clear light highlighting the glossy sheen of the chocolate and the textures of the meal

- **Cinematic lighting**: This type of lighting is used to create dramatic and visually appealing images, often for use in film or television. Here are some examples of keywords: *global illumination*, *cinematic light*, and *three-point lighting*:

Figure 6.38 – /imagine prompt - Cityscape photography of romantic couple kissing in Paris, Eiffel tower in background, cinematic light

- **Unreal Engine lighting**: This type of lighting is used to create realistic and immersive images, often for use in video games or other virtual environments. Here are some examples of keywords: *ray tracing*, *neon glow*, and *artificial lights*:

Figure 6.39 – /imagine prompt - A futuristic cityscape, bathed in the neon glow of artificial lights, in the style of street photography

- **Portrait lighting**: This is the primary light source in an image, and it is typically used to illuminate the subject's face. Here are some examples of keywords: *dim dark lighting, bright natural sunlight, light behind, dramatic lighting, high saturation, high contrast and very bright,* and *soft lighting*:

Figure 6.40 – /imagine prompt - the Four, oil, 2014, 76 x 72 in, in the style of realist still life photography, delicate markings, realist impressionism, flat shading, Portrait lighting, oil painting, traditional oil painting

- **Backlighting**: This type of lighting illuminates the subject from behind, which can create a sense of depth and mystery. Here are some examples of keywords: *rim lighting*, *backlighting*, and *realistic lighting*:

Figure 6.41 – /imagine prompt - beautiful dear running in a magical forest, wildlife photography, rim lighting, backlighting

- **Ambient lighting**: This type of lighting provides a general level of illumination in an image. Here are some examples of keywords: *ambient lighting, dramatic lighting, spotlight, studio lighting, warm colors, warm tone, fill lighting,* and *golden hour*:

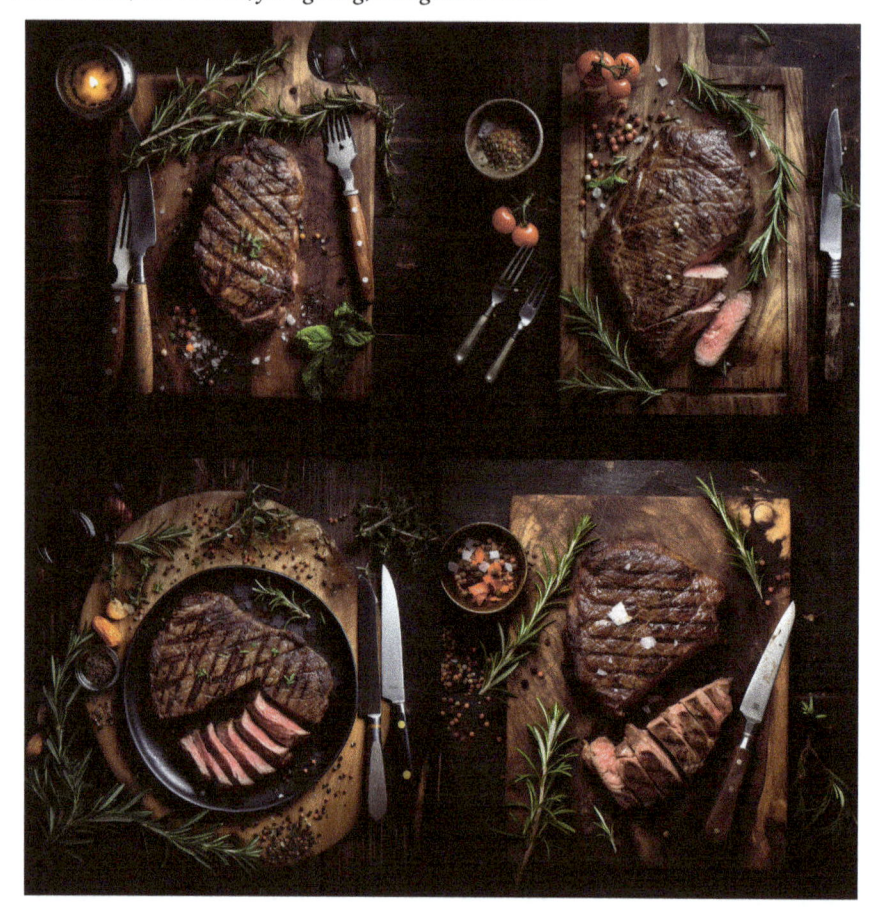

Figure 6.42 – /imagine prompt - steak grilling on top view plate on dark wood table and knife, in the style of gourmet food studio shot, refined aesthetic sensibility, food photography, shaped canvas, fine lines, delicate curves, pastoral charm, rtx, studio lighting

Camera angles

Camera angles also play an important role in composing an image in Midjourney and can be used to create different emotions and effects. You can use camera angles to guide the viewer's eye, create a sense of depth, and emphasize different elements of your image. Here is a list of some common camera angles and how they can be used:

- **High angle**: A high-angle shot looks down on the subject from above. This type of shot can make the subject look small and insignificant, or it can be used to show a sense of scale. For example, you could use a high-angle shot to show a large crowd of people or a vast landscape. To achieve a similar effect, you can also use *wide-angle, wide-angle shot, wide perspective, panoramic shot, ultra-wide angle, extreme wide shot, long shot, extreme long shot,* or *far shot*:

Figure 6.43 – /imagine prompt - A high angle shot of a city skyline at night

- **Low angle**: A low-angle shot looks up at the subject from below. This type of shot can make the subject look powerful or intimidating. It can also be used to create a sense of awe and wonder. For example, you could use a low-angle shot to show a tall skyscraper or a towering mountain. To achieve a similar effect, you can also use *knee-level shot, ground shot, ground-shot angle*, or *ground-level shot*:

172 Upgrading Your Prompt for Optimal Results

Figure 6.44 – /imagine prompt - A low angle shot of a skyscraper, making it look towering and impressive

- **Eye level**: An eye-level shot is taken from the same height as the subject's eyes. This type of shot is often used to create a sense of intimacy and connection with the viewer. It can also be used to show the subject's perspective. For example, you could use an eye-level shot to show a person walking down the street or a child playing in the park. To achieve a similar effect, you can also use *point-of-view shot*, *eye-level angle shot*, *straight-on shot*, *narrow-angle shot*, or *extreme narrow-angle shot*:

Figure 6.45 – /imagine prompt - An eye level shot of two people talking, creating a sense of intimacy and connection

- **Dutch angle**: A Dutch angle is a shot that is tilted on its axis. This type of shot can create a sense of unease or disorientation. It can also be used to add dynamism and interest to an image. For example, you could use a Dutch angle shot to show a person running away from danger or a scene of chaos and confusion. To achieve a similar effect, you can also use *full-body shot*, *establishing shot*, *medium shot*, *medium full shot*, *straight-on shot*, or *over-the-shoulder shot*:

Figure 6.46 – /imagine prompt - A Dutch angle shot of a vintage blue car chase, high-speed photography

- **Bird's-eye view**: A bird's-eye view is a shot that is taken from directly overhead. This type of shot can be used to show a large area or to show the relationship between different elements in a scene. To achieve a similar effect, you can also use *satellite view*, *aerial view*, *aerial perspective*, *drone footage*, *top-down shot*, *top-down perspective*, *elevated shot*, or *far-angle*:

Figure 6.47 – /imagine prompt - A bird's eye view of a forest, showing its vastness and complex landscape, nature photography, mist --ar 4:2

- **Back angle**: Also known as a reverse angle shot, this is a shot that is taken from behind the subject. This type of shot can be used to create a sense of mystery or suspense, as the viewer cannot see the subject's face. It can also be used to show the subject's body language or to highlight the background:

Figure 6.48 – /imagine prompt - A back angle shot of an astronaut walking on Mars, Astrophotography, in the style of NASA shot

In addition to these basic camera angles, you can play with more creative ones, such as *upside-down shot* or even *extreme narrow-angle shot*. You can also combine camera angles with other lighting and composition techniques to create even more stunning images. Experiment with different camera angles and types of shots, such as *extreme close-up*, *close-up*, *medium close-up*, *medium shot*, and *full*, and see what kind of results you get!

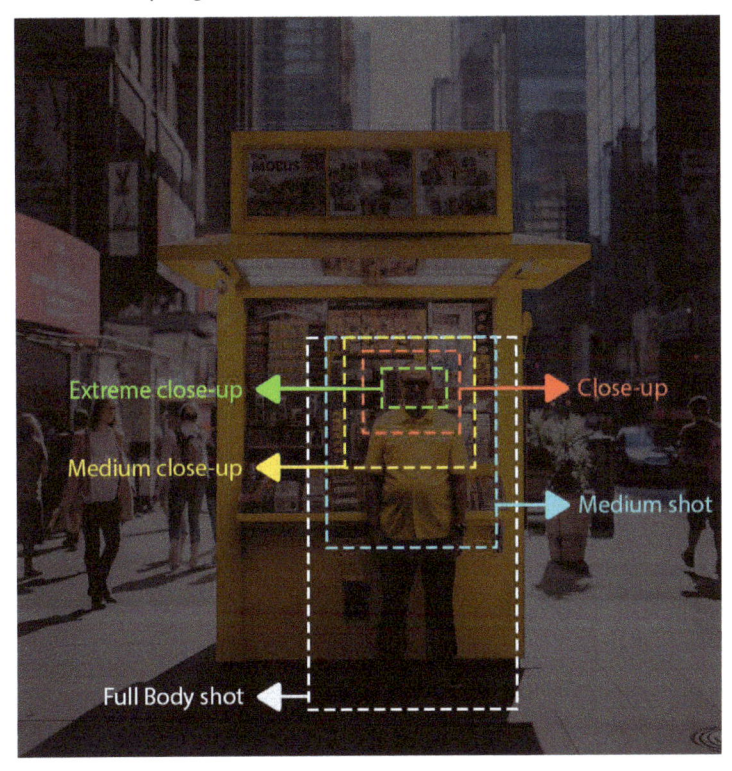

Figure 6.49 – Camera shot sizes

The styles and keywords highlighted here are just a starting point. The realm of artistic expression is boundless, and Midjourney offers a vast expanse of concepts to represent styles. If you ever need more inspiration, use the *describe* command. And when you're not sure about your prompt... well, just keep reading and you will learn about another command that can help you with your prompt.

The right words

Earlier in this chapter, we discussed a command that allowed us to add words to our chosen images. After that, we looked into various artistic, lighting, and camera styles that can enhance our images. Now, we're moving forward to learn how to combine these elements and identify the keywords needed for an effective prompt.

The /**shorten** command in Midjourney is truly interesting. For people like me, who started using Midjourney when there was little information on how to best use it and create effective prompts, we often wrote long descriptions. We thought that giving all the details was important, believing that Midjourney would understand and use all the specifics. However, over time, it became clear that giving the bot too much information often led to more complicated and sometimes less desired results. Midjourney might not consider much of the information in very long prompts. This is why the /**shorten** command is useful. It is a valuable feature, designed to optimize prompts by highlighting influential words and suggesting potential omissions.

Let's see how we can use it. For testing purposes, I'm going to share with you my first prompt ever… really, it's a relic particularly since this is from the time when we only had the version 2 model, and it's curiously large, so don't show it to anyone else:

/**imagine prompt** `A little bunny dressed as an astronaut floats in space, her helmet reflecting the vibrant stars of the Milky Way. She's on a mission to catch flying bananas, which are zooming past her in a rainbow of colors. Her spacesuit is adorned with bunny ears and a tail, and her jetpack is shaped like a giant banana. In the background, futuristic spaceships glide by, their engines glowing in the darkness. The bunny astronaut is determined to catch her bananas, but she's also enjoying the incredible view. She can see the whole galaxy spread out before her, and it's the most beautiful thing she's ever seen, vibrant colors, high detail, colors, rabbits, bananas with wings, hyper-realism cinematic, octane render, futurism, stylize, photorealistic, redshift render, 8K.`

Let's shorten it: start by typing the /**shorten** command in the chat box, as with any other command, and insert your prompt and hit *enter*:

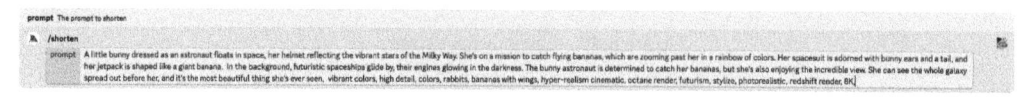

Figure 6.50 – How to use the shorten command

Now, a new message will appear with the review of your prompt, showing the important tokens, a numbered list of five shortened prompts, and the corresponding buttons so you can choose the ones you want to generate, and also a **Show Details** button (see *Figure 6.51*). In the **Important tokens** section, you can see that there are some words highlighted in bold, which means that Midjourney understands them as the most important tokens in your prompt, and other words that are struck through, meaning they don't have value. Let's hit the **Show Details** button, to understand a little more about the importance of our tokens and what we can learn with this:

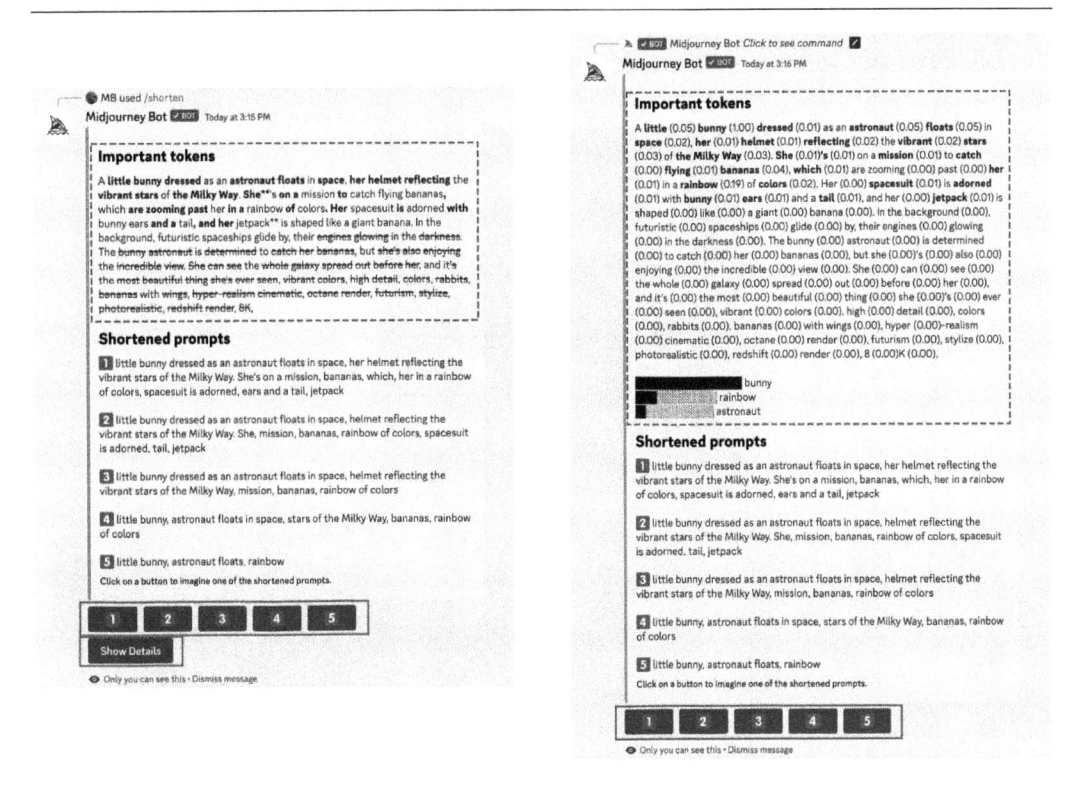

Figure 6.51 – List of the shortened prompts

> **Tokens**
>
> When using the **/shorten** command (keep in mind that it is not compatible with multi-prompts or the `--no` parameter), Midjourney assesses the strength of each token (word, phrase, or syllable) in your prompt, assigning them ratings from 0 to 1. This allows users to compare the effectiveness of individual tokens. While the scoring system provides insights, it's essential to note that even tokens with low scores can significantly influence the final image.

As you can see in the image on the right-hand side of *Figure 6.51*, most of the words appear with a rating of **0.00**, which means that they add little or nothing to the prompt and will certainly be ignored by Midjourney. In short, in a prompt of 124 words, only 27 have any value and only 3 are really relevant.

So, we can now understand the logic of our shortened prompts. The shortened versions retained the general aesthetic of the original prompt, but some nuances were lost when certain *ineffective words* were removed.

In the following figure, you can see the image generated with the original prompt and below that the suggestions for each of the short prompts:

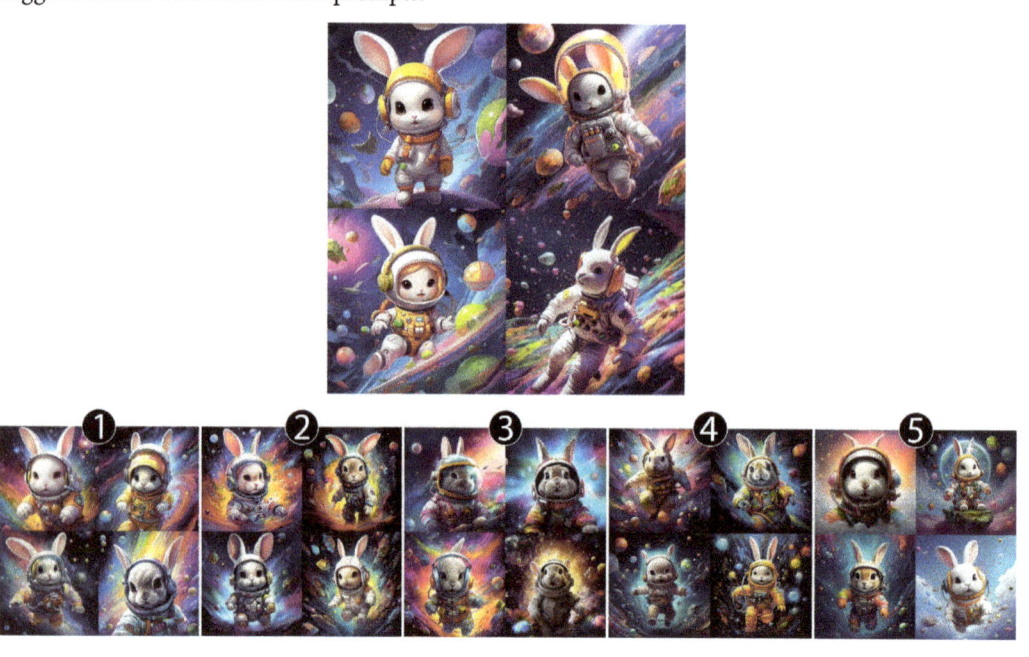

Figure 6.52 – Image generated with original prompt and with the shortened prompts

We can confirm that the results of the generated images are similar and the amount of information I had initially was useless. In the end, we can see that in none of the prompts was the final result as expected. At the time, what I wanted was an image showing a rabbit in space catching flying bananas. Although the concept is strange, it would have been possible if the prompt had been constructed more coherently from the start. We can't expect these commands to work miracles. In essence, the **/shorten** command serves as a guide, helping us understand how Midjourney interprets prompts. It offers a clearer perspective on which words are impactful and which might be redundant. However, the final decision on prompt construction remains in our hands, ensuring that the desired aesthetic and details are captured.

Figure 6.53 – Bonus image using /imagine prompt - astronaut bunny catching bananas in space, surrounded by bananas

After understanding the intricacies of the **/shorten** command and its significance in creating effective prompts for Midjourney, it's essential to explore another dimension that can further elevate our experience. This brings us to the collaboration between Midjourney and ChatGPT, two powerhouses in their respective domains. The fusion of their capabilities opens up a world of possibilities for users.

The synergy between ChatGPT and Midjourney is undeniable. ChatGPT, a powerful AI tool with vast knowledge and conversational abilities, can be a valuable asset for Midjourney users. By feeding ChatGPT with information about Midjourney and its prompts, you can enhance its power to generate creative and effective prompts.

For those who are unfamiliar, ChatGPT is like a knowledgeable friend who can answer almost any question. Midjourney, on the other hand, is a platform that turns your imagination into stunning visual realities. When combined, these two tools can produce remarkable results.

For example, you can provide ChatGPT with examples of effective Midjourney prompts and then request ChatGPT to create a new prompt based on a given concept. The result? Customized, actionable prompts that enhance the Midjourney experience.

180   Upgrading Your Prompt for Optimal Results

Moreover, the integration of ChatGPT with Midjourney goes beyond just prompt generation. There are resources available, such as the **Midjourney Prompt Generator** (**MGPM**) on GitHub, which further streamlines the process. This tool configures ChatGPT to generate Midjourney prompts, offering users a quick and efficient way to produce high-quality prompts. It's perfect to get new word ideas and new concepts, and if like me you're not fluent in English, this tool can be an excellent ally.

If you are interested in using ChatGPT to generate Midjourney prompts, you can visit www.openai.com, go to **Log in**, and enter your account details, or create a new one by clicking on **Try ChatGPT**:

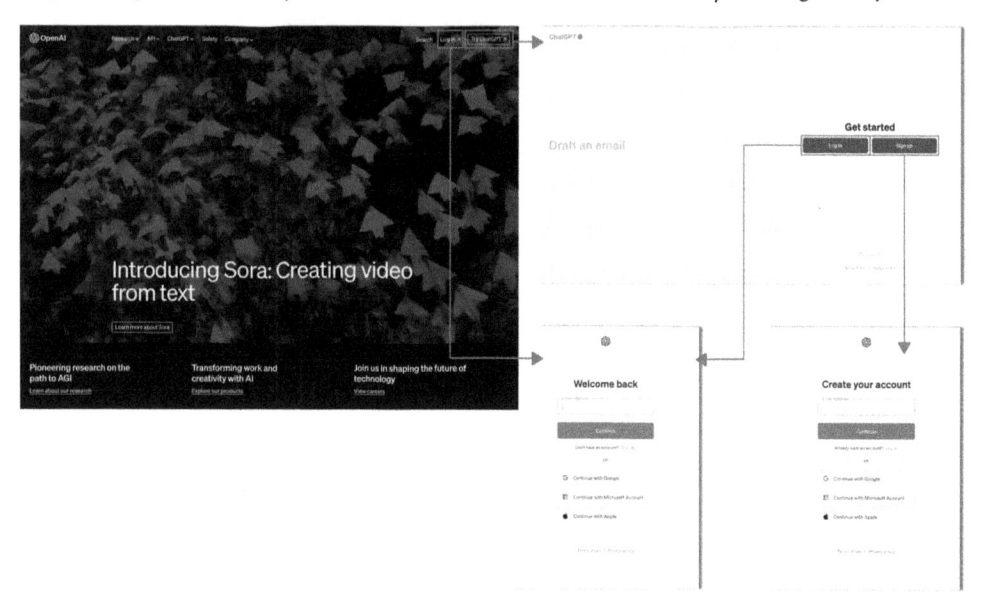

Figure 6.54 – OpenAI login page

After that, you will be able to start talking with your new AI friend. Now, use the following link to access the MPGM: https://github.com/chatgpt-prompts/ChatGPT-Midjourney-Prompt-Generator?ref=allthings.how

Click on the midjourney-prompt-generator-prompt-for-chatgpt file and a new page will open:

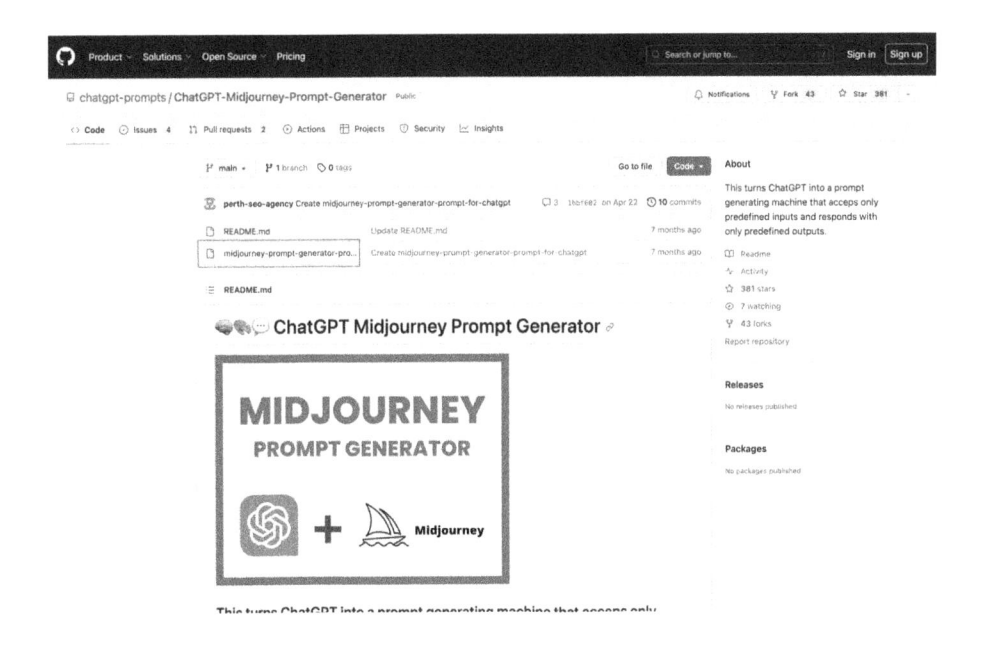

Figure 6.55 – MPGM GitHub

Next, click on the **Copy raw contents** button:

Figure 6.56 – Copy the raw contents from MPGM GitHub

Now go to www.chat.openai.com and start a new chat. Paste the contents you had copied from the previous link and press *enter*. ChatGPT will now send you a message saying **Midjourney Prompt Generator Mode ready**.

You can now start using it. Type the [Start MPGM] command in the chatbox and send it. Next, type [prompt] followed by what you want ChatGPT to give you a prompt for, for example, [prompt] extreme close shot of an angel. Click **Send** and ChatGPT will then return three scenarios for you to choose from and more instructions. In any case, I advise you to go back to the GitHub page and there you'll find the author's instructions about all the available commands and how to end the MPGM. Note that you can only use the MPGM in the chat that you pasted the *Copy raw contents*.

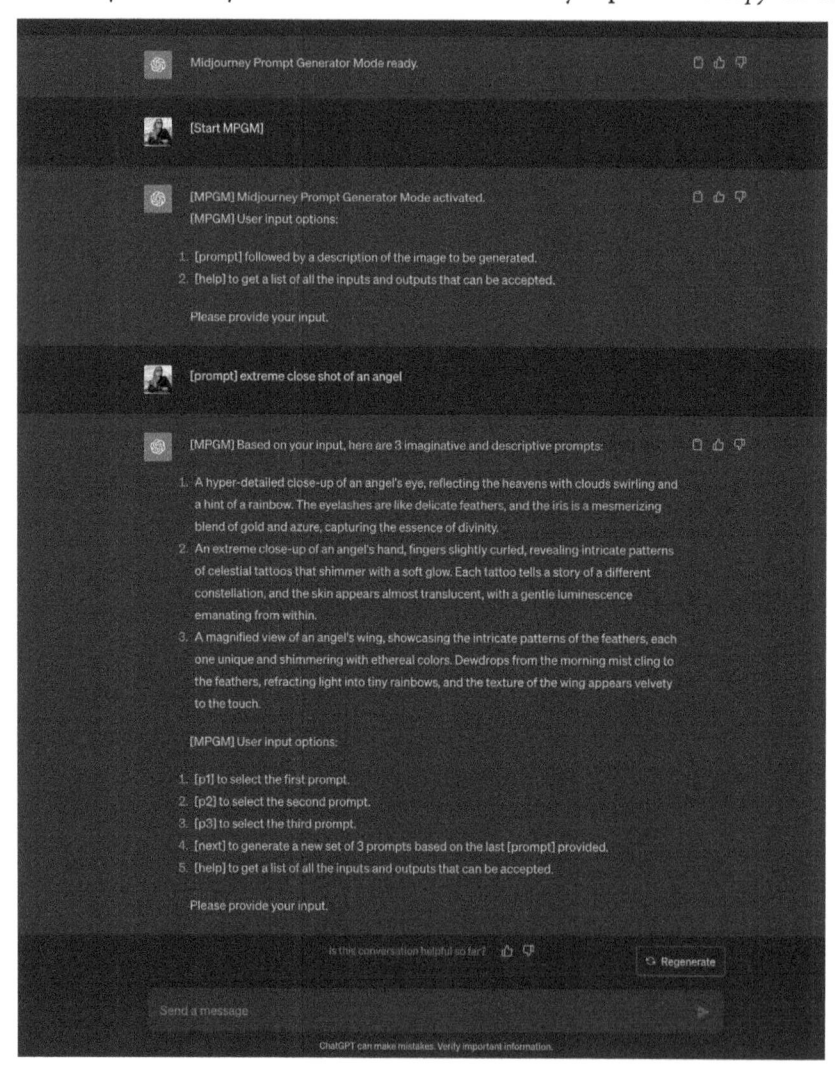

Figure 6.57 – By using MPGM, you always get three prompts based on your initial idea

Overall, all of these tools and commands are a valuable asset for any Midjourney user who wants to create more effective and stunning images. By combining the **/shorten** command, ChatGPT, and the MGPM, users can push the boundaries of their creativity and produce more effective and stunning images.

Summary

In this chapter, we explored the depths of Midjourney, mastering the foundational *describe* command and discovering the diverse artistic, photographic, and lighting styles that can enhance our creations. We also cracked the code of the *shorten* command, uncovering its power to optimize prompts and highlight essential words. Additionally, we have forged a partnership between ChatGPT and Midjourney, leveraging their combined expertise to generate creative and effective prompts.

In the next chapter, we will explore how to customize Midjourney and go even further with our image generation.

7
Customizing Midjourney – Settings, Preferences, and Unleashing Creative Prompts

With a solid foundation in navigating Midjourney's capabilities, we will shift focus in this chapter to personalize your experience with the tool. As we've acquainted ourselves with the basics of generating art with Midjourney, it's time to dive into the nuances of its settings and preferences. This chapter is designed to guide you through customizing Midjourney, allowing you to optimize your use of this powerful tool to match your individual creative vision.

In this chapter, we will cover the following topics:

- **An overview of Midjourney's settings**: In this section, you will get a comprehensive understanding of Midjourney's various settings and learn how they influence the art creation process.
- **Customizing your preferences**: Here, we will Learn to adjust and personalize your Midjourney experience, crafting custom codes for values you often use in your prompts.
- **Fine-tuning our images with the Style Tuner**: In this section, we explore the **Style Tuner** to tailor the visual style of your Midjourney images. This section will show you how to use the /**tune** command to generate different visual styles based on your prompts, helping you choose the best fit for your artistic direction.

By the end of this chapter, you will have gained essential skills in understanding and applying each setting in Midjourney for enhanced control over your art. You will know how to create personalized code for frequent values in your prompts, make your workflow more efficient, and use the Style Tuner to customize the appearance of your images to add a unique flair to your Midjourney projects.

Technical requirements

To make the most of this chapter and follow along with the hands-on activities, you'll need to ensure you've set up everything we covered in *Chapter 2*. Here's a quick recap:

- **Midjourney account**: You'll need an active Midjourney account. If you haven't registered yet, refer to *Chapter 2* for a step-by-step guide.

- **Discord installation**: Ensure you have Discord installed on your device, as we'll be working with it alongside Midjourney. Again, the installation guide is available in *Chapter 2*.

- **Payment plans**: At the time of writing, Midjourney has no free or trial plan. A Midjourney paid plan is necessary. Ensure that you've selected the most suitable plan for your needs.

- **Server creation**: Having your own server can significantly enhance and expedite your experience with Midjourney. If you've yet to set up a server in Discord, *Chapter 2* covers the process in detail.

An overview of Midjourney's settings

One of the key features of Midjourney is its simple but powerful settings menu, which gives users a high degree of control over the creative process.

In this chapter, we will explore the various settings available in Midjourney and explain how they can be used to achieve different results. We will also provide tips on how to customize your preferences to create the perfect Midjourney experience. Start by simply typing /**settings** in the message box:

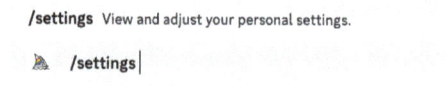

Figure 7.1 – Type the /settings command in the message box and hit enter

A list of your current settings will appear. Understanding and utilizing these settings can significantly enhance the quality and specificity of the generated images:

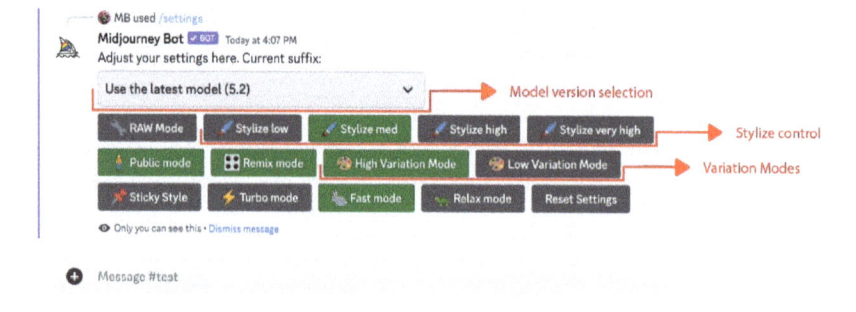

Figure 7.2 – A list of my settings

The settings in Midjourney control the fundamental aspects of your image generations. These settings include the following:

- **Model version selection**: This setting allows you to select the version of the Midjourney AI model that you want to use. Choose from various model versions, including V1 to V5.2, and anime-focused models such as Niji V4, Niji V5, and Niji V6. Each version has its own unique strengths and weaknesses (*Chapter 3* covers all the model versions). By default, Midjourney uses the latest model.

- **The style raw parameter**: Only available with model versions 5.1, 5.2, and 6.0, the `--style raw` parameter allows for a less stylized Midjourney default aesthetic, producing more raw and unfiltered image outputs:

Figure 7.3 – The same /imagine prompt - surreal sparkling delight with a touch of gold sweetness, with raw in the left grid and without raw on the right grid

- **Stylize control**: The `--stylize` or `--s` flag adjusts the level of artistic interpretation. Lower values yield images truer to the prompt, while higher values enhance artistic abstraction. As we saw earlier in *Chapter 4*, the stylization values can be set in a variety of ways. They can be entered manually, or they can be set using one of the four stylize buttons (**Stylize low**, **Stylize med**, **Stylize high**, or **Stylize very high**) in the settings:

Figure 7.4 – The same /imagine prompt - surreal sparkling delight with a touch of gold sweetness, with the different stylize values and raw style also active

The **Stylize low** button corresponds to a value of 50 (equivalent to adding `--s 50` at the end of your prompt). The **Stylize med** button is the default value and corresponds to a value of 100 (`--s 100`). The **Stylize high** button corresponds to a value of 250 (`--s 250`). The **Stylize very high** button corresponds to a value of 750 (`--s 750`):

Figure 7.5 – The same /imagine prompt - surreal sparkling delight with a touch of gold sweetness, with the different stylize values

- **Public/Stealth mode**: This setting in Midjourney determines the visibility of your image generations. When set to *Public* your images are accessible and can be viewed by anyone using Midjourney. Conversely, in *Stealth* mode, your images remain private and are only visible to you. However, this privacy is contingent on using a private server. It's important to note that if you're on your own server but have set your mode to *Public*, all your creations become public and are visible to others on the Midjourney site. Similarly, if you're in *Stealth* mode but are using a public server, your images will still be visible to other users. This highlights the importance of understanding how server choice interacts with your selected mode to control the privacy of your creations. Note that *Stealth* mode is only available for subscribers to the *Pro* plan. For more information about this, visit www.midjourney.com/account.

- **Remix mode**: This is a very useful setting that enables you to change prompts, parameters, model versions, or aspect ratios between variations. This means you can start with an initial image and then tweak these elements to generate new variations. When using *Remix*, the new images are influenced by the general composition of the starting image. This is particularly useful for evolving a subject, changing settings or lighting, or achieving complex compositions. Another way to activate and deactivate *Remix* mode is by using the **/prefer remix** command.

Once *Remix* is enabled, the variation buttons (**V1**, **V2**, **V3**, and **V4**) under image grids change behavior. You can select an image and choose **Make Variations** to start the remixing process. The variation buttons turn green instead of blue, indicating that you can modify the prompt for each variation:

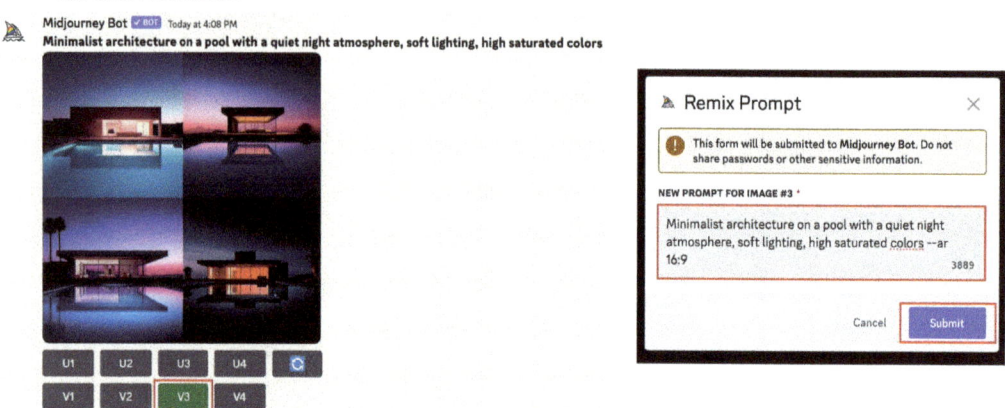

Figure 7.6 – After choosing image three to remix, I just added a new aspect ratio to be generated in a new grid of images; in this case, I added to the initial prompt, --ar 16:9

During each variation, *Remix* allows you to edit your prompt. This editing can include changing the subject, style, or other parameters. (Note that only parameters that normally influence variations will work in *Remix* mode. These include *aspect ratio*, *chaos*, *image weight*, *quality*, *seed*, *stop*, *stylize*, *tile*, and *video*.)

In my case, I added to the end of my prompt a different aspect ratio, `--ar 16:9`, and hit **Submit**. Based on the third image, Midjourney generated four new images with a new, more horizontal aspect ratio:

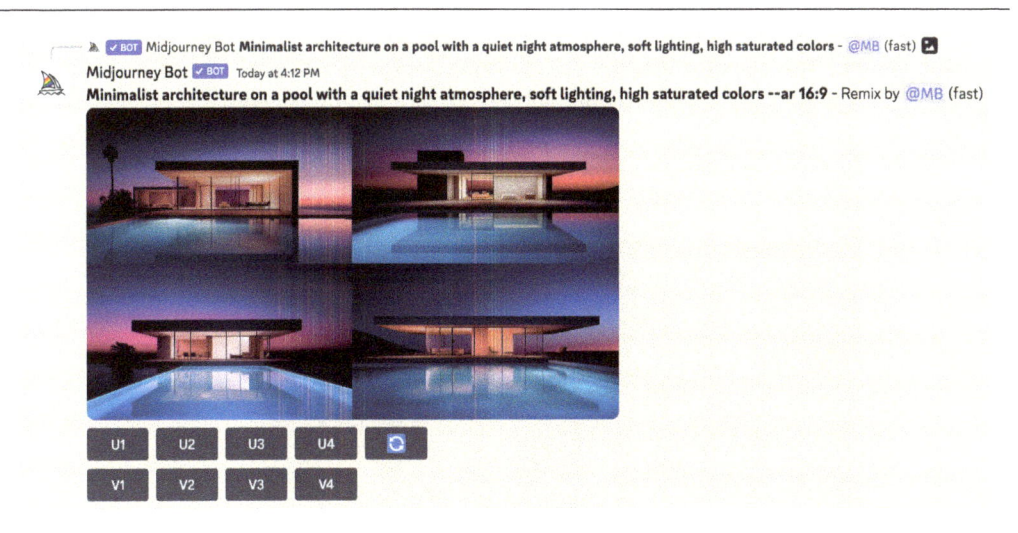

Figure 7.7 – A new grid of images generated with --ar 16:9

The *Remix* feature extends its versatility to upscaled images as well. To illustrate this, let's upscale *image one* with the new aspect ratio, 16:9, and click on the **Vary (Region)** button. A pop-up window will appear, inviting you to select the *Lasso* tool. Once selected, trace an area on the image that you wish to modify, and then type in the desired element:

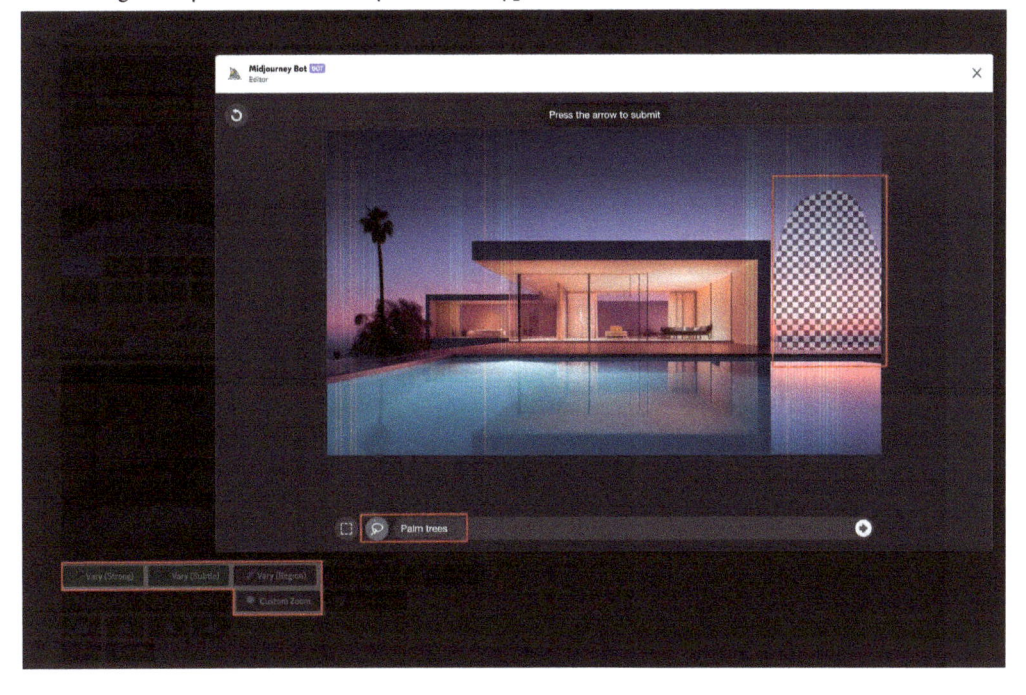

Figure 7.8 – An example of the Remix mode with the Vary (Region) option

Alternatively, you can refine the prompt by utilizing the **Vary (Strong)** or **Vary (Subtle)** option, or even employing the **Custom Zoom** button (we explored this in *Chapter 3* in the *My first prompt – now what?* section).

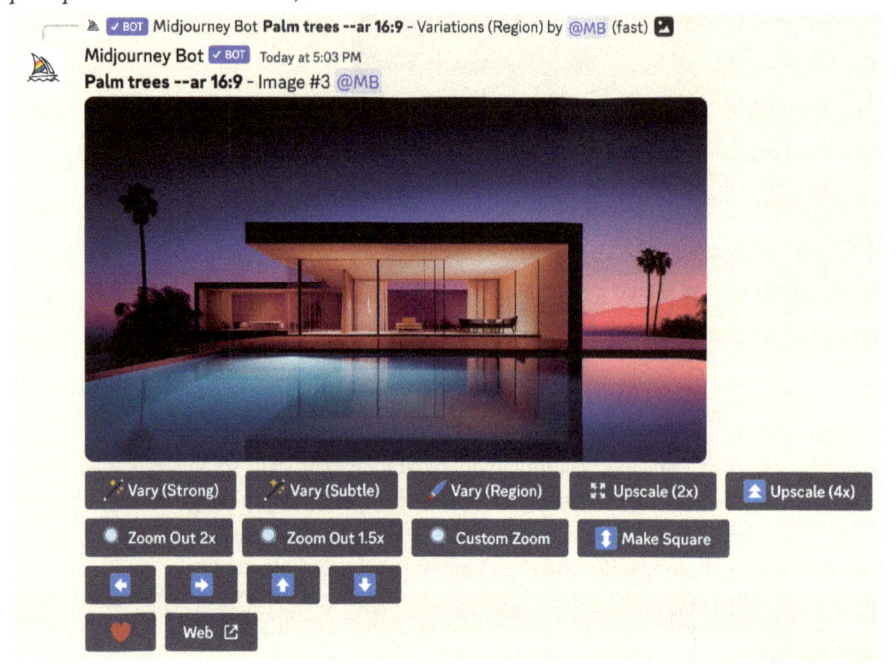

Figure 7.9 – My final result with my new palm trees also on the right side

- **High Variation Mode and Low Variation Mode**: These settings offer us the flexibility to control the degree of change in the AI-generated images. **High Variation Mode** is ideal for generating a wide range of concepts from a single image. It introduces significant changes to the composition, elements, colors, and details within the image. This means that when using this mode, the **V1, V2, V3**, and **V4** buttons under each image grid, or the **Vary (Strong)** button under an upscaled image, will produce new images with considerable differences from the original. **High Variation Mode** is the default setting and is the one I used in all the images generated in this book.

    Conversely, **Low Variation Mode** is designed for making subtle adjustments or refinements to an image. It retains the main composition of the original image but introduces minor changes to its details. Selecting this mode and then using the variation buttons, or the **Vary (Subtle)** button under an upscaled image, results in variations that are closely aligned with the original image but with nuanced differences.

Using the previous prompt, let's see the difference in results using **Low Variation Mode**. I will start by activating **Low Variation Mode** in my settings and then write an **/imagine prompt** `Minimalist architecture on a pool with a quiet night atmosphere, soft lighting, high saturated colors --ar 16:9`, and I will choose *image four* to try out some variations:

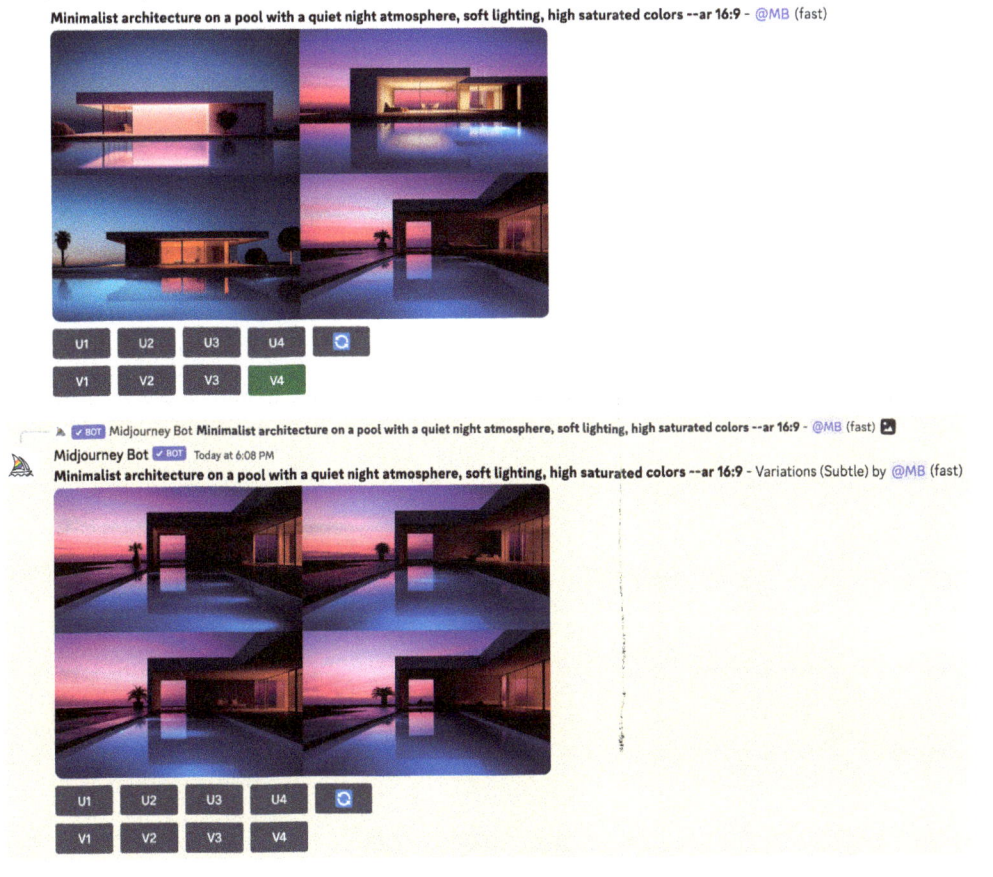

Figure 7.10 – The new grid of variations' image is very close to the original image with only subtle variations

As you can see, the variations that are generated next are very subtle and similar to the image I chose. This is the advantage of using **Low Variation Mode**. Now, let's engage in a bit more creative exploration by leveraging *Remix* mode, which is still active. Let's revisit the first grid of generated images. Click the **V4** button once more to modify the prompt slightly. We'll introduce candles and yellow furniture by the pool, seeking an ambiance of tranquil night, soft lighting, and high-saturated colors. Type in the following:

/**imagine prompt** Minimalist architecture on a pool, with modern yellow furniture, candles floating in the pool, with a quiet night atmosphere, soft lighting, high saturated colors --ar 16:9:

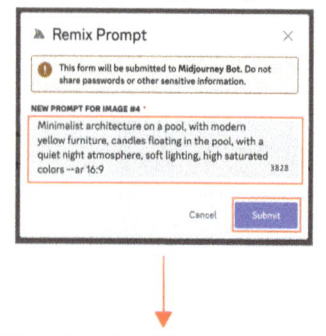

Figure 7.11 – The new grid of variations' image with my yellow furniture

As you can observe, employing **Low Variation Mode** enabled me to achieve the subtle variations I desired within the same image (although I miss the candles floating in the pool, but I can still try to include them later with the **Vary Region** feature). **Low Variation Mode** proves exceptionally valuable when you're content with the overall concept of your image but seek to refine specific aspects or delve into subtle variations. To fast-toggle between the High and Low Variation Modes, you can use the **/prefer variability** command:

An overview of Midjourney's settings   195

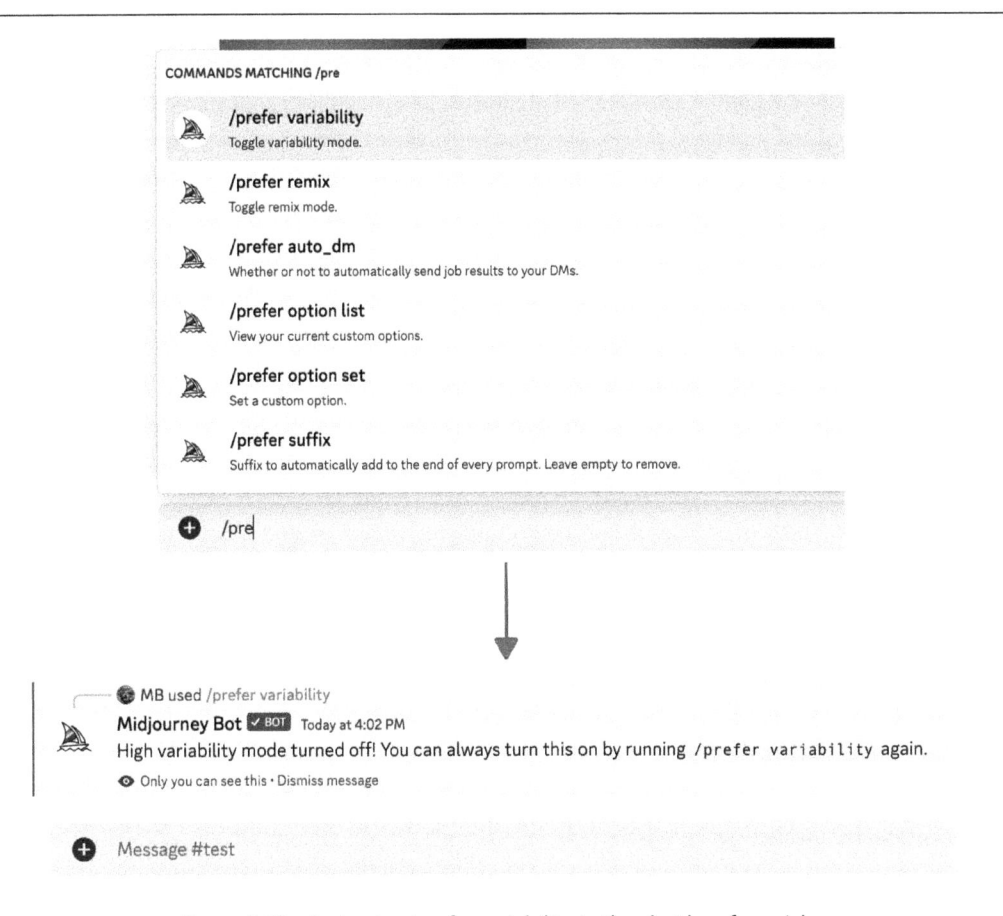

Figure 7.12 – Just enter /prefer variability in the chat box for quick toggling between the two variation modes

- **Fast/Turbo/Relax mode**: This was previously discussed in *Chapter 4*. Midjourney utilizes powerful **graphics processing units** (**GPUs**) to interpret and process each prompt, with different subscription plans offering varying amounts of monthly GPU time (for more information about this, visit www.midjourney.com/account). This GPU time is primarily used in *Fast* mode. To switch between modes, just type the **/fast**, **/turbo**, or **/relax** command in the chat box, or use the `--relax`, `--fast`, or `--turbo` parameter at the end of a prompt for a single job:

    - **Fast mode**: Known for instant GPU access, *Fast* mode aims to provide instant access to a GPU. The average job in Midjourney takes about one minute of GPU time. Factors such as upscaling, aspect ratios, and model versions can affect this time.

- **Relax mode**: Known for unlimited image generation, this is available for all plan subscribers, allowing the creation of an unlimited number of images without consuming GPU time. Jobs are queued and processed as GPUs become available. Wait times can vary but generally range from 0–10 minutes per job. Priority in the queue depends on how much the system has been used and resets with each subscription renewal. A limitation of this mode is that certain features such as permutation prompts and the legacy upscaler are not available in *Relax* mode.

- **Turbo mode**: Known for expedited image generation, this mode uses a high-speed experimental GPU pool for extremely quick image generation. Jobs in *Turbo* mode generate up to four times faster but consume twice as many subscription GPU minutes. This mode is available only with Midjourney model versions 5, 5.1, and 5.2. If Turbo mode is selected but GPUs are unavailable or incompatible with the selected model version, the job will run in *Fast* mode.

- **Reset settings**: This setting allows you to go back to all the default settings.

- **Sticky Style**: This setting helps the process of applying your preferred style code generated with the Style Tuner (we will learn more about this in the *Style Tuner* section) to future prompts. By enabling **Sticky Style** using the **/settings** command, you can save the last `--style` parameter used in your personal suffix (there's more on this next in the *Customizing your preferences* section). This eliminates the need to repeatedly enter the same style code for subsequent prompts.

  To switch to a different style code, simply write `--style` followed by your new code or deselect **Sticky Style**. **Sticky Style** acts as a convenient shortcut, saving you time.

Having explored the fundamental settings that govern the Midjourney experience, we've witnessed their impact on the creative process. Now, let's embark on a deeper exploration of personalization. In the forthcoming section, we'll uncover the art of tailoring Midjourney to your needs.

Customizing your preferences

Midjourney allows you to create custom options using **/prefer** commands. This feature enables you to add commonly used parameters, styles, or concepts automatically to the end of prompts, saving time and ensuring consistency. Imagine you use a lot of these terms in this sequence: *realistic, breathtaking, cinematic, hyper-realistic, photorealistic and incredibly detailed, professional lighting, studio lighting, 100mm, photography gallery, winning photography*. Instead of writing this in every prompt, you can create a shortcut for it. Let's learn how!

/prefer option

The **/prefer option** command lets you create or manage custom options. To view your current custom options, simply type **/prefer option list** in the chat box and hit *enter*. I have utilized this feature to create a custom option named `wallpaper`, which is just one of up to 20 custom options that a user can have:

Figure 7.13 – An example of my custom options on the right

By using the **/prefer option set** `<option> <value>` command, I was able to define my custom option. In the `option` field, I entered the word `wallpaper`, and in the `value` field, I specified an aspect ratio of `16:9` and a high stylization setting of `250`.

A crucial part of this customization is ensuring that the generated images are free from any text or letter-like elements, which can be distracting or irrelevant to the desired visual output. To achieve this, I discovered that using the `--no` parameter followed by `newspaper text` is particularly effective. The term `newspaper text` here acts as a filter within Midjourney, instructing the AI to avoid generating images with elements that resemble text, especially in the style commonly found in newspapers. This approach has become an invaluable part of my workflow, allowing me to produce images that closely align with my artistic vision, free from unnecessary textual elements.

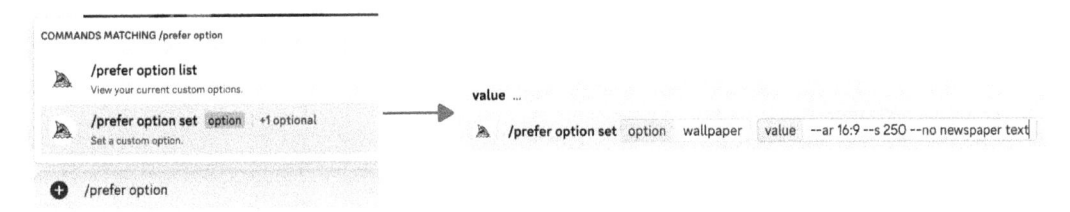

Figure 7.14 – An example of my /prefer option set command

This command proves invaluable for swiftly incorporating multiple parameters (*aspect ratio*, *chaos*, *no*, *quality*, *repeat*, *seeds*, *stop*, *style*, *stylize*, *tile*, *version*, *video*, and *weird*; check out *Chapter 4* to learn more about each one) into your prompts, saving you time and effort. Now, let's see how it works. I will create a new prompt, and at the end of it, I will simply add my new parameter:

/imagine prompt `futuristic car, streamlined design, designed in a realistic and photographic style, morning light --wallpaper:`

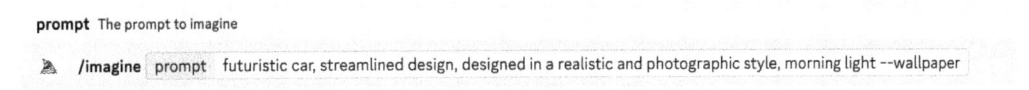

Figure 7.15 – My prompt with the parameter I previously created

Automatically, Midjourney interprets my `--wallpaper` parameter as **/imagine prompt** `futuristic car, streamlined design, designed in a realistic and photographic style, morning light --ar 16:9 --s 250 --no newspaper text`:

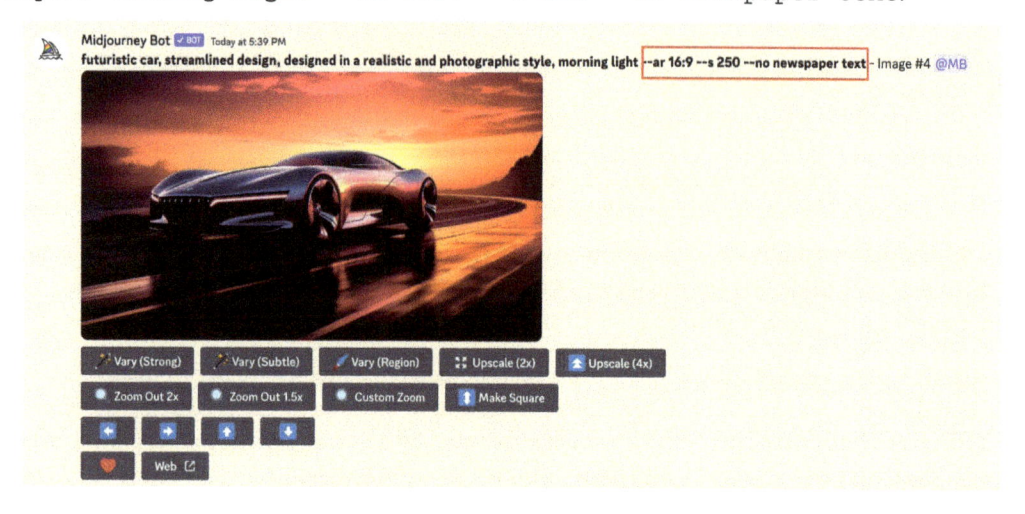

Figure 7.16 – An example of the correct interpretation that Midjourney made of my parameter

If you need to delete a custom option, type **/prefer option set**, hit *enter*, write the name of your option (don't use the value field), and hit *enter* again:

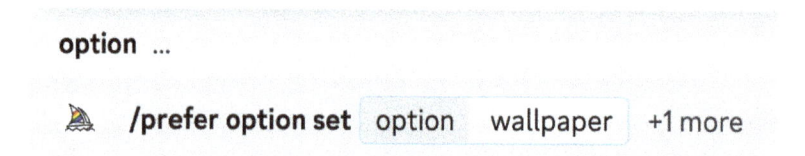

Figure 7.17 – Deleting a custom option

/prefer suffix

The **/prefer suffix** command automatically appends a specified suffix to all your prompts. This is particularly useful for parameters you want to be constant in all your prompts, and you can add more than one parameter to **/prefer suffix**:

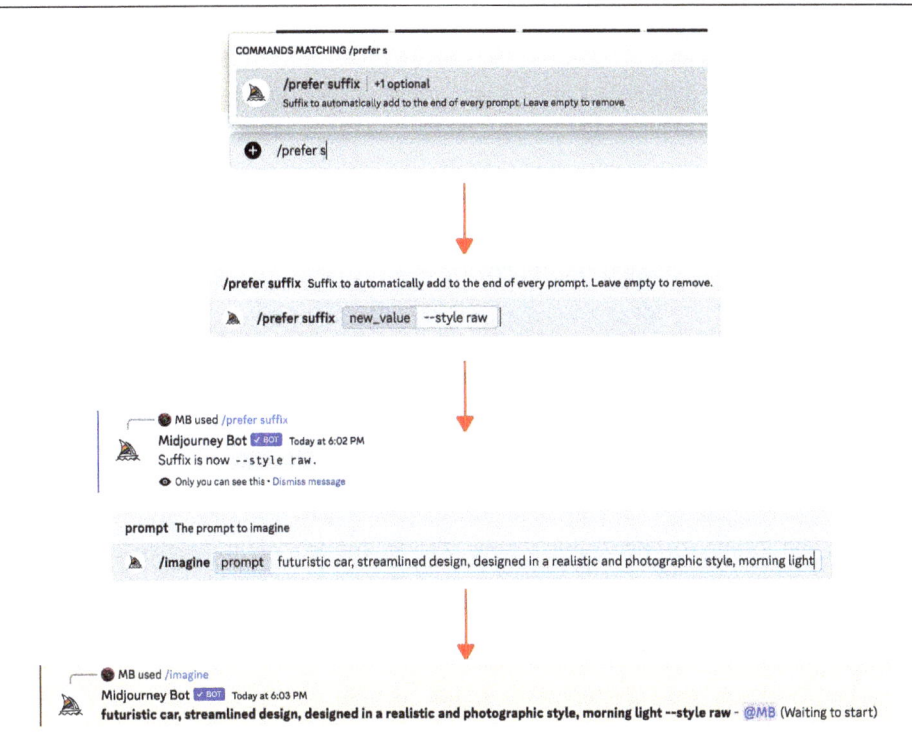

Figure 7.18 – How to create your own suffix to automatically be added to all your prompts

In my case, I want to start using `--style raw` in all my prompts. I will start by simply typing **/prefer suffix** and enter `--style raw` in the `new_value` field. Note that only parameters (*aspect ratio, chaos, no, quality, repeat, seeds, stop, style, stylize, tile, version, video,* and *weird*; check out *Chapter 4* to learn more about each one) can be used with **/prefer suffix**:

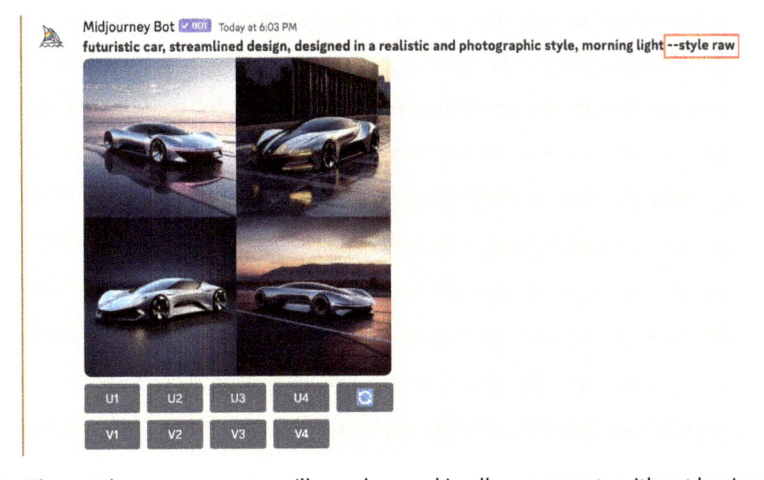

Figure 7.19 – The --style raw parameter will now be used in all my prompts without having to write it

As before, as I write my prompt and hit *enter*, Midjourney again automatically generates my images in `--style raw`. To clear a preferred suffix, you can reset the settings in the **/settings** command or just write **/prefer suffix** in the message box and hit *enter*, and the suffix will be removed:

MB used /prefer suffix

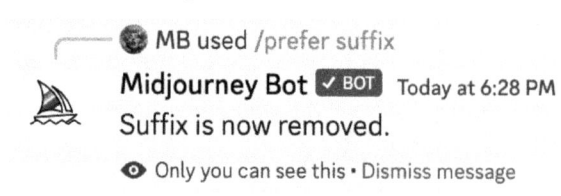

Midjourney Bot ✓ BOT Today at 6:28 PM
Suffix is now removed.

👁 Only you can see this • Dismiss message

Figure 7.20 – How to remove a prefer suffix

These custom preferences offer flexibility and efficiency in the creative process, allowing you to focus more on the artistic aspect and the story you want to tell with your image, rather than repetitive typing of parameters. By mastering these settings and custom preferences, you can significantly enhance your experience with Midjourney, making it more efficient and tailored to your specific needs. As you move forward, the introduction of the Style Tuner opens up a new dimension of personal artistic style, enabling you to further refine and personalize your visual creations.

Fine-tuning our images with the Style Tuner

Style Tuner is a game-changing feature, offering a unique way to fine-tune the aesthetic nuances of your images. The Style Tuner provides a hands-on approach to visual styling, allowing you to experiment with various visual styles and select the one that best aligns with your artistic vision.

Only available while in *Fast* mode, this feature lets you personalize the appearance of your images by using the **/tune** command, generating a range of sample images that showcase different visual styles, based on your prompt. This process not only aids in visualizing the potential outcomes but also empowers you to choose your favorite style.

The journey begins with the **/tune** command. Typing **/tune** followed by a prompt, such as **/tune prompt** `cute kitten with pink flowers`, sets the stage for customization:

prompt The base prompt to use for the tuner

/tune prompt cute kitten with pink flowers

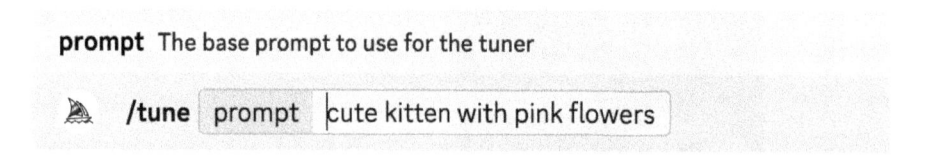

Figure 7.21 – Type /tune followed by your prompt and hit enter

After entering your prompt, you will see a message from Midjourney, with two drop-down buttons. Here, you can choose the style direction (**16**, **32**, **64**, or **128**) and the mode (**Default** or **Raw**). When you select a style direction in Midjourney, it determines how many different style variations Midjourney will generate for you. The number you choose is actually doubled by Midjourney to create a series of 2 x 2 grid images. For instance, if you pick 32 styles, Midjourney will generate 64 images (32 times 2). This process is resource-intensive, which explains why it can be considered expensive in terms of GPU credits. Remember that if you need to know how many *fast hours* you still have, just type **/info** in the message box.

The cost of creating a new Style Tuner in Midjourney, based on your chosen number of style directions, is measured in *fast hours*. Here's a breakdown of using the *Standard* subscription plan, which includes 15 hours per month in *Fast* mode:

- 16 styles require 0.15 *fast hours*, which is about 9 minutes
- 32 styles need 0.3 *fast hours*, or approximately 18 minutes
- 64 styles take 0.6 *fast hours*, equaling 36 minutes
- 128 styles use 1.2 *fast hours*, which is around 72 minutes

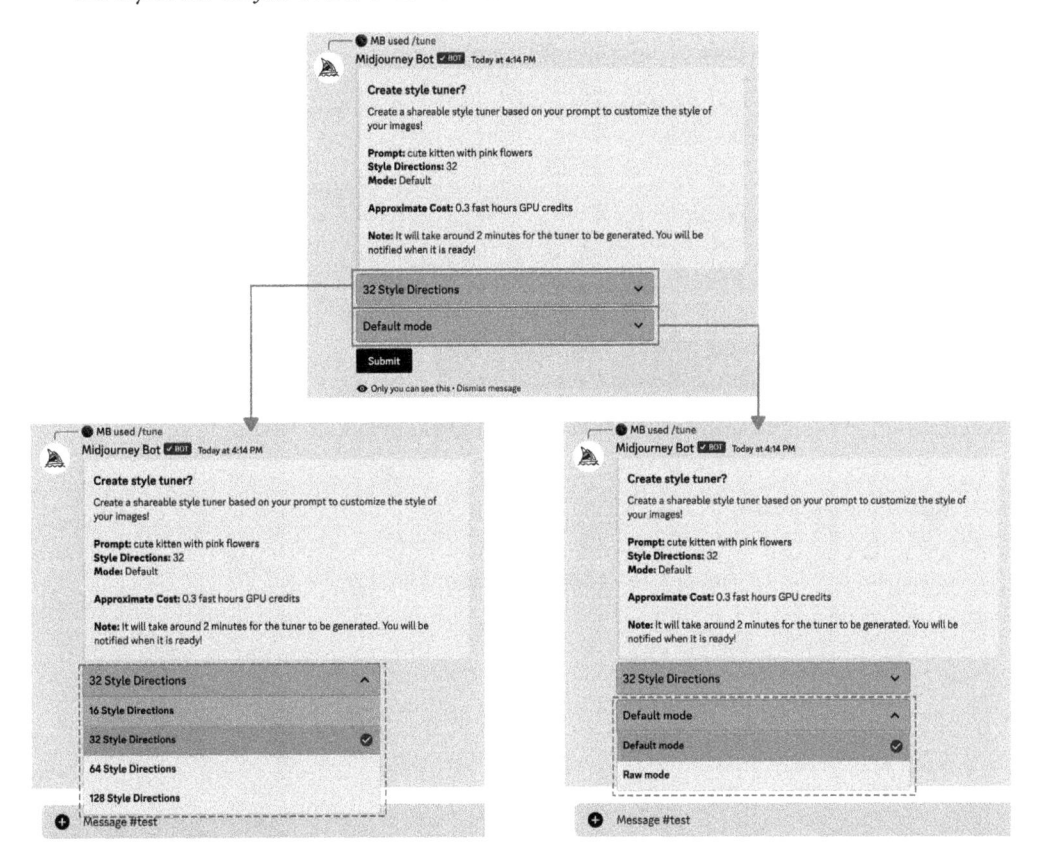

Figure 7.22 – Choose the number of image pairs and the mode

In essence, the Style Direction number you select impacts the variety and quantity of the style variations Midjourney will produce, as well as the computational time needed to create them. In my case, as I don't usually use `--style raw`, I will choose **Default** mode, and I want to have 32 pairs of images to compare. Then, I will click **Submit**, following which a message asking for my confirmation will appear:

> **MB used /tune**
>
> **Midjourney Bot** ✓ BOT   Today at 4:14 PM
> (edited)
>
> **Create style tuner?**
>
> Create a shareable style tuner based on your prompt to customize the style of your images!
>
> **Prompt:** cute kitten with pink flowers
> **Style Directions:** 32
> **Mode:** Default
>
> **Approximate Cost:** 0.3 fast hours GPU credits
>
> **Note:** It will take around 2 minutes for the tuner to be generated. You will be notified when it is ready!
>
> [32 Style Directions ⌄]
>
> [Default mode ⌄]
>
> [Are you sure? (Cost: 0.3 fast hrs GPU credits)]
>
> 👁 Only you can see this • Dismiss message

➕ Message #test

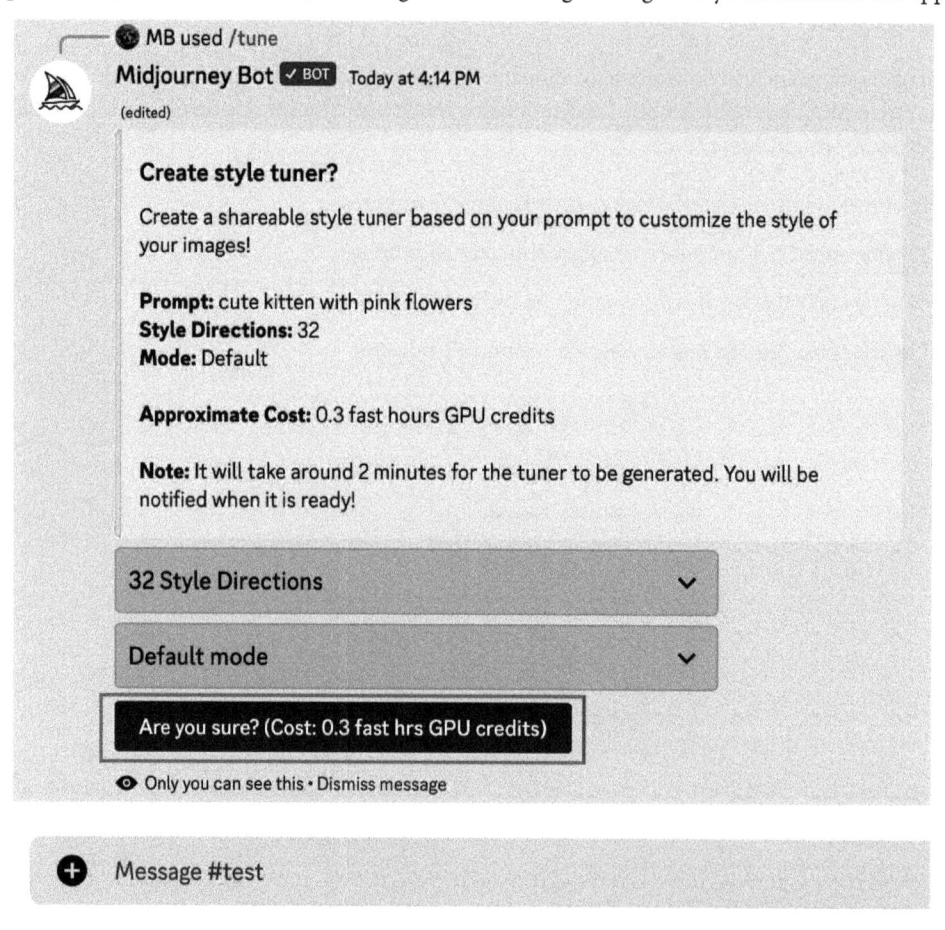

Figure 7.23 – Confirming your submission

After submitting and confirming your job, the Midjourney bot will send you a message, telling you that 64 out of your 64 jobs are being processed. In my case, it is 64 because I chose **32 Styles Directions**, which will generate 64 images; as we saw before, if I had chosen 16, we would have 32 images, and if I had chosen 128 directions, we would have 256 images. Right after this, you will get a new message, saying that your Style Tuner is ready, along with a unique URL that will direct you to your Style Tuner page:

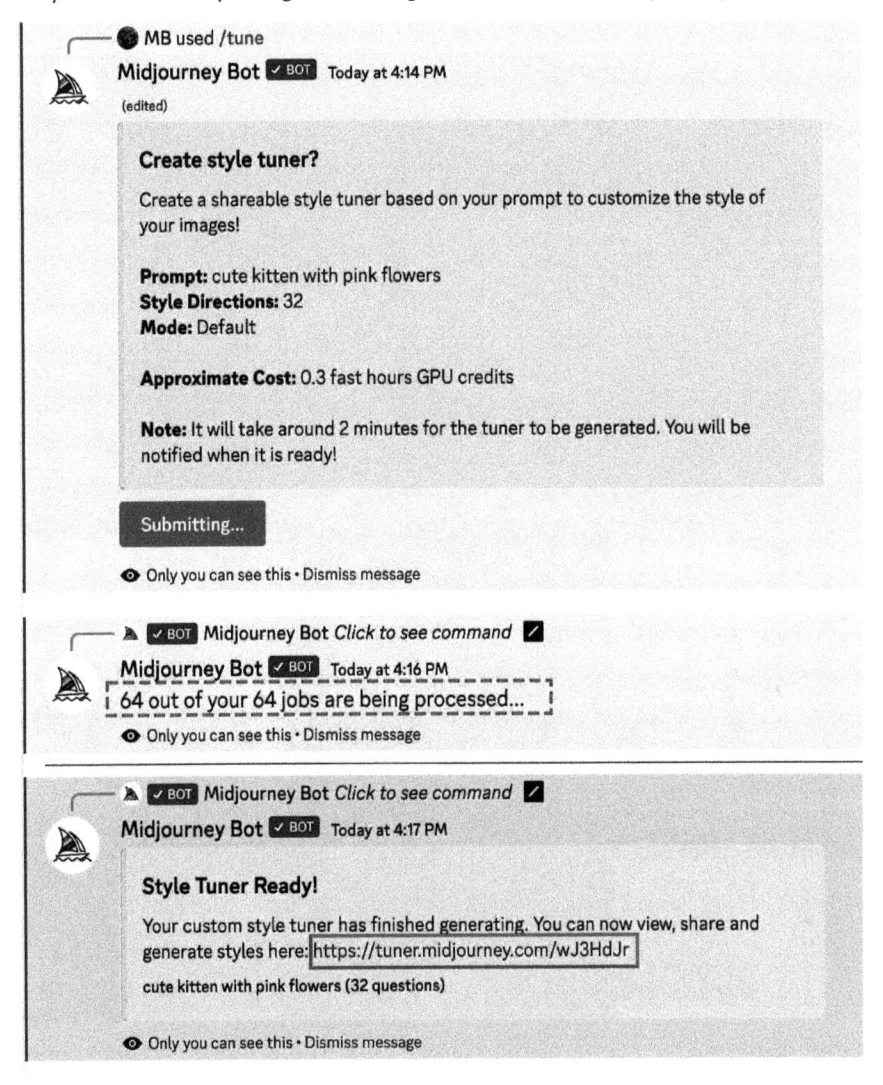

Figure 7.24 – Click on the link and start tuning your style

Once you click on this URL, you will be presented with a variety of styles in pairs of images. By selecting images from the pairs, you can determine the visual direction for your prompt. The recommended approach is to select 5–10 styles that resonate most with you, refining the aesthetic direction of your prompt. If you don't really like either of the images in that row, leave the middle box empty. You can choose the way you pick your images – two styles at a time (my favorite view is with one grid on the left and another grid on the right) or in a grid:

Figure 7.25 – You can choose the way you pick your images – two styles at a time (my favorite) or in a grid

Once you've made your selection, Midjourney provides a unique code, which you can use to consistently apply this customized look to your future creations. Click on the icon to copy your prompt plus the style code; in my case, this style is `--style 3hPQAKop37r`. Alternatively, just go to the bottom of the page to get your code.

The personalized style code is then used in your prompts, such as **/imagine prompt** `a lion in a jungle --style 3hPQAKop37r`, allowing Midjourney to adopt your customized style. In my case, I think I found the perfect style to make illustrations for a children's book – what do you think?! Check out a selection of images that resulted in this style code here: `https://tuner.midjourney.com/code/3hPQAKop37r`.

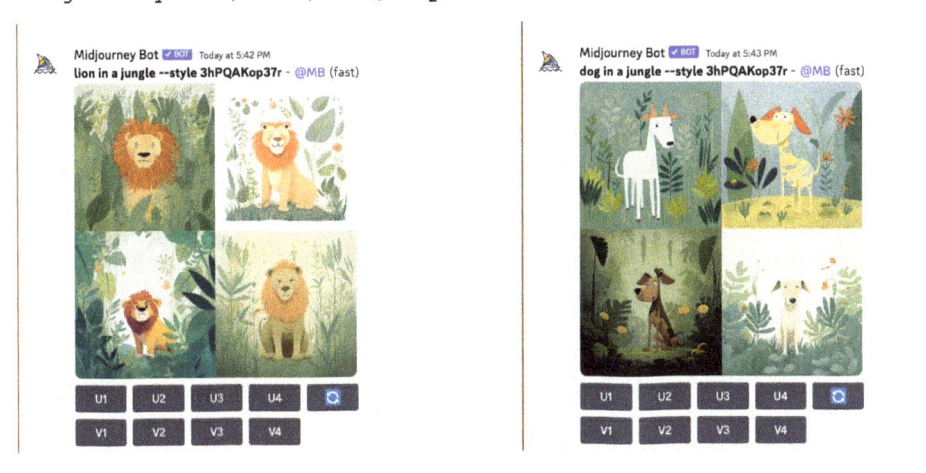

Figure 7.26 – Using my new style in my prompts

As the Style Tuner is very versatile, it enables the creation of numerous style codes from just one session. By selecting different images, you can generate entirely new style codes. This means you can return to the previous URL of your Style Tuner page (in my case, I used `https://tuner.midjourney.com/wJ3HdJr`), select another group of images, and get a new unique style code to use. This way, you don't need to use more *fast hours*, and another benefit is that you can share this page with your friends, and they also can generate new style codes without any cost.

However, there are a lot of different things you can do with style codes.

Combining style codes

By using `--style <code1-code2>`, you can merge different styles, opening up new avenues for creativity. For example, I created another style, `--style 7aqdgiJ606ZIYQ`, which was based on the `psychedelic flowers` prompt (you can visit my Style Tuner page at `https://tuner.midjourney.com/kqrv7UK` and create your own codes), so you can imagine that the results are very different from the one we have been testing based on the cats prompt. Let's check how the two codes behave together. I'll start by writing the same prompt with the two styles – **/imagine prompt** `dog in a jungle --style 3hPQAKop37r-7aqdgiJ6O6ZIYQ`:

Figure 7.27 – Combining style codes

As you can see, now I have my dog and lion illustrations with a bolder graphic look, perfect to make a small animation and create a video to promote my children's book:

Figure 7.28 – The same prompt I used before but now with the two styles combined – /imagine prompt - dog in a jungle --style 3hPQAKop37r-7aqdgiJ6O6ZIYQ and / imagine prompt - Lion in a jungle --style 3hPQAKop37r-7aqdgiJ6O6ZIYQ

Adjusting with --stylize and --raw

To fine-tune the impact of your style code, the `--raw` and `--stylize` parameters can be altered, enhancing the desired effect. To illustrate this, let's use this prompt with the different parameters. Again, I'll start by writing the same prompt with the two styles, followed by the *stylize* value I want – **/imagine prompt** `dog in a jungle --style 3hPQAKop37r-7aqdgiJ6O6ZIYQ`:

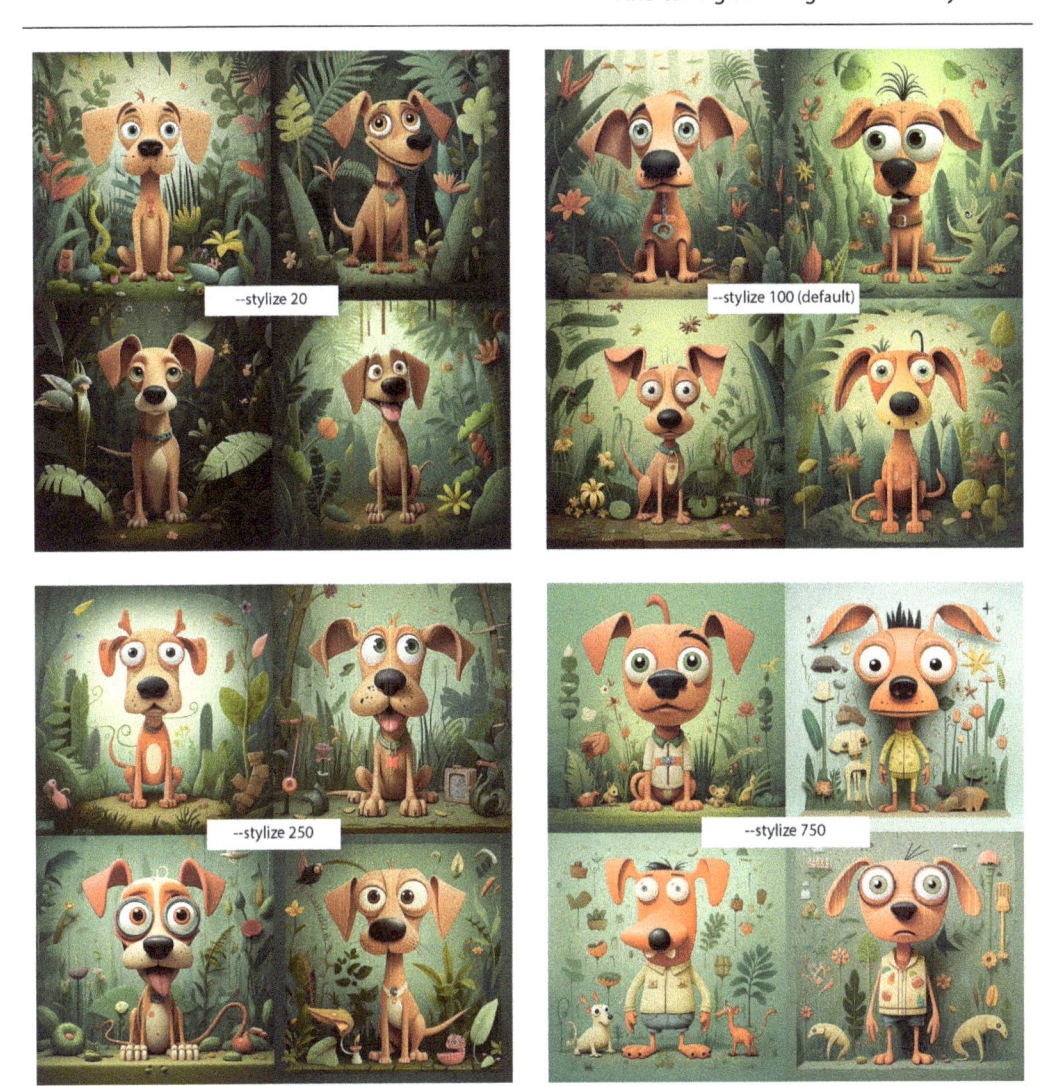

Figure 7.29 – The same prompt I used before with the two styles combined and the stylized value – /imagine prompt - dog in a jungle --style 3hPQAKop37r-7aqdgiJ6O6ZIYQ --stylize <enter value>

Now, let's do the same but just with `style raw`, using **/imagine prompt** `dog in a jungle -style raw-3hPQAKop37r-7aqdgiJ6O6ZIYQ`:

Figure 7.30 – The same prompt I used before with the two styles combined and style raw – /imagine prompt - dog in a jungle --style raw-3hPQAKop37r-7aqdgiJ6O6ZIYQ

Experimenting with random styles

For those who enjoy surprises, the `--style random` parameter generates by default a random 32 visual/style directions style code, offering a fresh perspective each time. You can also use this parameter followed by other number of visual/style directions, such as `--style random-16`, `--style random-64`, or `--style random-128`, and Midjourney will automatically generate a random code. Here are some examples that I got with these styles:

Figure 7.31 – Using random codes from other lengths of tuners

The `--style random` allows us to explore a range of artistic styles by specifying both the number of possible visual directions and the percentage of then we want to have in our images. This percentage, which can range from 1% to 100%, essentially controls how many of the available style elements are utilized by the AI in the image generation process.

For instance, the `--style random-16-9` command means that out of 16 available style directions, the AI will incorporate elements from approximately 9% of these directions. This doesn't limit our choices; rather, it guides the AI in selecting a subset of styles to apply, creating a specific blend of artistic influences in the final image. Similarly, `--style random-128-90` suggests that the AI will use elements from 90% of the available 128 style directions, offering a broader and more diverse stylistic range in the generated image.

It's crucial to understand that this percentage influences the variety and distinctiveness of the styles in the output. A lower percentage, such as `--style random-64-5`, means the AI selects from a smaller pool of style elements, leading to images with more distinct and varied styles. Conversely, a higher percentage, such as `--style random-64-100`, involves a larger selection of style elements, potentially resulting in subtler differences between styles as more elements are blended together.

Remember, as we learned at the beginning of this chapter, that you can always save and reuse your codes by clicking on the **Sticky Style** button or using the `prefer` suffix to automatically add your preferred style to every prompt. Alternatively, you can create various `prefer` options for each of your codes.

> **Style Tuner technical details**
>
> The Style Tuner is not compatible with image prompts that do not include a text prompt. Style Tuner is only available in *Fast* mode. Additionally, the Style Tuner is at the time of writing exclusively compatible with model version 5.2. Style Tuner extends its versatility by supporting additional parameters such as `--aspect`, `--chaos`, and `--tile`, as well as multi-prompting.
>
> However, these parameters only influence the sample images you see in the Style Tuner page. They don't alter `--style <code>` produced by it. Therefore, adjusting these parameters won't impact the results of any future Midjourney images created using `--style <code>`.
>
> Every style you create or modify with the Style Tuner is assigned a unique code. To view or share a specific style, you can directly access its dedicated page on the Style Tuner website. Simply append the unique style code to the URL: `https://tuner.midjourney.com/code/StyleCodeHere`. This direct access feature makes it easy to revisit, refine, or share your custom styles with others.

As a fun fact, if you really like one of the images that appear on your Style Tuner page, just know that you can import it into your Discord and upscale, edit, or reroll it, or any other option you want.

To do this, while on your Style Tuner page, choose the **Pick your favorites from a grid** option, right-click on the image you want, and choose **Open Image in New Tab**:

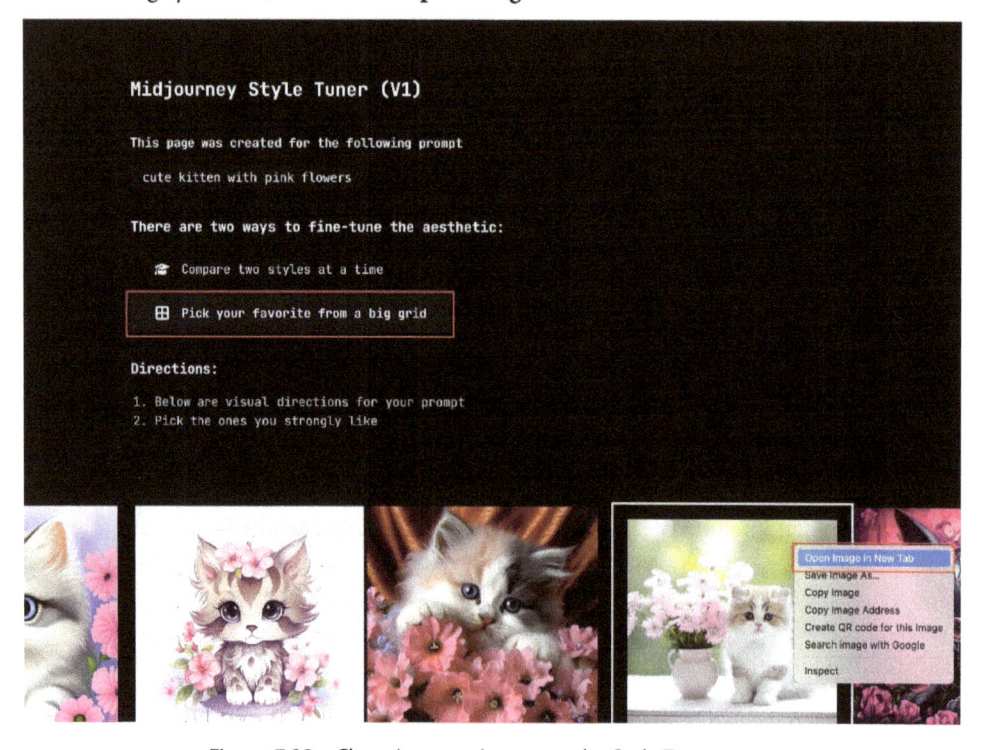

Figure 7.32 – Choosing your image on the Style Tuner page

In the new tab, go to the URL and copy the numbers and characters that represent the Midjourney job ID of the image:

cdn.midjourney.com/cb3a5fe8-c93b-499c-89a2-cc6e99acf8f6/0_0_640_N.webp?method=shortest

Figure 7.33 – Copying the numbers and characters

Now, back in your Discord server, go to your message box and type /**show**. The job ID will appear. Paste the numbers and characters you have copied and press *enter*:

/**show** job_id cb3a5fe8-c93b-499c-89a2-cc6e99acf8f6

Figure 7.34 – Paste the numbers and characters into the Job ID and press enter

After this, you will have a grid of images, including the image you chose on your Style Tuner page, and you can now use it like any other image generated on Midjourney.

Figure 7.35 – The usual grid of images – in this case, with my preferred cat image from the Style Tuner page

Summary

In this chapter, we journeyed through the intricate settings and customization options of Midjourney, equipping you with the knowledge to fine-tune your creative process. From mastering the nuances of various settings to exploring the innovative Style Tuner, you're now adept at tailoring Midjourney to your vision. These features and options not only enhance your efficiency but also deepen your engagement with Midjourney, allowing your creativity to flourish without technical constraints.

In *Chapter 8, Exploring Practical Use Cases and Pushing Boundaries*, we will shift our focus from the technical mastery of Midjourney to its practical and professional applications in the real world. This chapter is designed for advanced users who are ready to go beyond the basics and explore storytelling and visual communication through AI art.

Part 4: Prompting for the Real World

The final part of the book extends the exploration of AI art creation into practical, real-world applications. It illustrates how Midjourney can be applied not just for personal projects but also within professional environments and across different industries. This section delivers insights into practical use cases, tips, and tricks, and wraps up with reflections on the journey with Midjourney, highlighting the transformative influence of AI on creativity and design.

This part has the following chapters:

- *Chapter 8, Exploring Practical Use Cases and Pushing Boundaries*
- *Chapter 9, Unlocking Tips and Tricks and Special Functions*
- *Chapter 10, Conclusion – Your Journey with Midjourney*

8
Exploring Practical Use Cases and Pushing Boundaries

Welcome to the most practical chapter of this book!

In this chapter, we will learn how to adapt and tailor our creations to meet our professional needs using **Midjourney**. The journey through this chapter is designed to showcase the versatile applications of Midjourney, guiding you through the process of creating immersive narratives and captivating moodboards and demonstrating the practical and professional applications of this tool.

In this chapter, we will cover the following topics:

- **Generating ideas and captivating moodboards**: In this section, we will see how we can use Midjourney as a creative partner to generate ideas and create compelling moodboards. This skill is crucial for visual storytelling and project planning, providing a foundation for creative projects.

- **The visual power of storytelling**: This section will show how you can use the knowledge you have acquired to create engaging stories. You will learn to craft narratives that evoke emotions, engage readers, and leave a lasting impression.

- **Creating brand sets with Midjourney – Icons and logos**: Here, we will understand the process of creating brand sets with Midjourney, including logos, and design simple elements such as **iconography**. This skill is vital for branding, identity design, graphic design, and visual communication.

- **Utilizing Midjourney's photorealism for product mockups**: Here, we will learn to create tailored mockups that seamlessly showcase your products, enhancing marketing campaigns and product presentations. Master the basics of achieving realistic outputs, such as food and interior photography, with Midjourney. You will also explore techniques for generating realistic and engaging mockups. These skills are beneficial to marketing, advertising, and product design teams as well to photographers and visual artists.

This chapter is structured to provide advanced insights into using Midjourney as a tool for creative exploration and professional application. Each section is crafted to empower you to push the boundaries of your creative work. By the end of this chapter, you will have mastered the use of Midjourney to optimize your workflows and save time, applying AI tools to enhance your professional projects efficiently.

Technical requirements

To make the most of this chapter and follow along with the hands-on activities, you'll need to ensure you've set up everything we covered in *Chapter 2*. Here's a quick recap:

- **Midjourney account**: You'll need an active account on Midjourney. If you haven't registered yet, refer back to *Chapter 2* for a step-by-step guide.
- **Discord installation**: Ensure you have Discord installed on your device, as we'll be working with it alongside Midjourney. Again, the installation guide is available in *Chapter 2*.
- **Payment plans**: At the moment, Midjourney has no free or trial plan. A Midjourney paid plan is necessary. Please ensure you've selected the most suitable plan for your needs.
- **Server creation**: Having your own server can significantly enhance and expedite your experience with Midjourney. If you're yet to set up a server in Discord, *Chapter 2* covers the process in detail.

Generating ideas and captivating moodboards

Midjourney's versatility extends beyond its ability to generate realistic images; it's also a highly effective tool for ideation and moodboard creation. In the fast-paced environment of a design or marketing agency, where client demands can vary widely, Midjourney emerges as an invaluable creative partner. Whether clients have a clear vision or are seeking direction, Midjourney can be an ally in visualizing various paths and possibilities.

For clients who know what they want, Midjourney can help in refining and expanding upon their initial ideas. It allows us, creatives, to present multiple visual directions, offering clients a tangible representation of potential outcomes. This can be particularly crucial in decision-making processes, where visual cues often speak louder than words.

Conversely, for those who are unsure of what creative direction they want to follow, Midjourney serves as a source of inspiration and exploration. It can spark fresh ideas, open up unexpected creative avenues, and provide a range of visual representations that guide the creative process. This aspect of Midjourney is especially beneficial in brainstorming sessions, where it can serve as a catalyst for generating innovative concepts and exploring unconventional approaches.

The power of Midjourney lies in its ability to transform abstract concepts into tangible visual representations. By providing clear and concise prompts, you can elicit from Midjourney images that embody the essence of your ideas. This process is invaluable in both guiding clients through

Generating ideas and captivating moodboards 217

the creative journey and aiding your creative mind to explore and present a variety of aesthetic and conceptual options.

Figure 8.1 – Example of a color moodboard generated in Midjourney

Moodboards are essential tools for bringing visual coherence to creative projects. They serve as a visual representation of the project's overall aesthetic, atmosphere, and tone. By combining Midjourney-generated imagery with existing visual references, you can create compelling moodboards that capture the essence of your ideas.

Midjourney's ability to synthesize diverse styles and imagery allows for the creation of unique and evocative moodboards. You can experiment with different color palettes, textures, and composition styles to create a visually cohesive representation of your project's vision.

Start by guiding Midjourney's creativity with specific prompts, for example, `moodboard for a dystopian future` or `moodboard for a sustainable food delivery service`. Also, don't be afraid to experiment with Midjourney's versatility in styles and mediums. From realistic to abstract art, and across painting, drawing, photography, and 3D modeling, Midjourney can help you discover your creative voice for any work, generating visuals closely aligned with your vision.

Over time, I found that the following steps were the right thing to do in order to avoid entering the rabbit hole of AI generation when creating project moodboards:

- **Start with a clear theme**: Define the moodboard's purpose and the emotions or ideas you wish to evoke, and input terms such as *moodboard*, *fashion*, *jewelry*, and *summer look* in your prompt
- **Guide with descriptive prompts**: Use specific prompts to direct Midjourney's output, focusing on desired colors, textures, and styles, for example, *moodboard*, *bright colors*, *luxurious*, and *in the style of haute couture*

- **Include diverse imagery**: Incorporate a variety of terms for visual elements that can evoke the specific look or emotion you want to convey, for example, *inspiring*, *pinterest*, *image showcase*, *product photography*, and *collage*

Let's see some practical use cases of this in terms of creative exploration:

Business and marketing strategies

For businesses strategizing their next big marketing campaign, Midjourney can be used to visualize concepts and themes. By inputting keywords related to the campaign's goals, Midjourney can generate a series of images that capture the essence of the proposed strategy, providing a visual foundation for further development. Here's a scenario:

- **Possible client briefing**: Design a campaign for our international product launch event for an eco-sustainable tech brand.

    Prompt example for moodboard: /imagine prompt `moodboard for a futuristic eco-friendly product launch event, showcasing innovative green technology and sustainable design elements`:

Figure 8.2 – Example of a moodboard for a futuristic eco-friendly product launch event

Event planning and celebrations

Event planners can utilize Midjourney to conceptualize themes for weddings, corporate events, or celebrations. By creating moodboards that reflect desired color schemes, themes, and atmospheres, planners can present clients with a visual representation of their event, facilitating decision making and ensuring alignment of vision. Here's an example:

- **Possible client briefing**: Conceptualize/Plan our wedding, keeping in mind the following – summer wedding, soft colors, and flowers.

    Prompt example for moodboard: **/imagine prompt** `moodboard for a summer garden wedding, featuring pastel colors, floral arrangements, and elegant outdoor settings`:

Figure 8.3 – Example of a wedding moodboard

Interior design projects

Interior designers can also use Midjourney to explore different design aesthetics, color palettes, materials, and layouts. By generating images that reflect various interior styles, designers can create moodboards that help clients visualize potential designs for their spaces. Here's an example:

- **Possible client briefing**: Remodel our living room, keeping in mind our preference for classic yet modern styles.

  Prompt example for moodboard: /imagine prompt `moodboard for modern minimalist living room interior, emphasizing neutral tones, natural light, and sleek furniture designs`:

Figure 8.4 – Example of a moodboard for interior design

Conceptualizing brand identities

In branding projects, Midjourney can be instrumental in developing brand identities. Designers can input elements of a brand's ethos and values to generate visuals that resonate with the brand's identity, aiding in the creation of logos, packaging, and overall brand aesthetics. Here's an example:

- **Possible client briefing**: Develop branding elements for our artisanal coffee shop.

    Prompt example for moodboard: **/imagine prompt** `moodboard logo and branding elements for a gourmet coffee shop, focusing on artisanal vibes, warm earthy colors, and a cozy atmosphere`:

Figure 8.5 – Example of a moodboard of colors and brand elements

Creative brainstorming

I have to say that this is one of my favorites in Midjourney, with the possibility of creativity and ideation being pushed forward with Midjourney's help.

Midjourney can serve as a brainstorming partner, which is perfect for when you or a creative team need fresh ideas. By inputting abstract concepts or themes, it can generate a range of visuals that spark new ideas and directions for projects, whether it's for advertising, content creation, or just personal artistic endeavors. Here is an example:

- **Prompt example for a brainstorm moodboard:** **/imagine prompt** `inspiration collage moodboard, material samples, bright colors, minimalist, future tech, pinterest --s 250`:

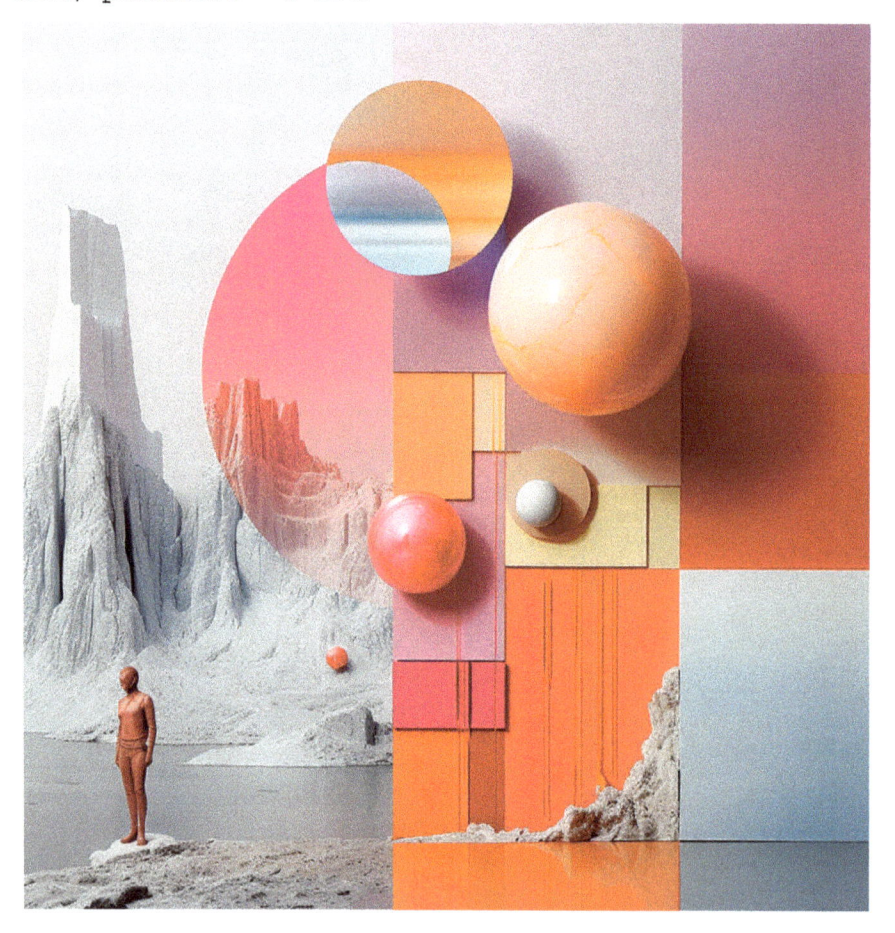

Figure 8.6 – Inspiring moodboard

In essence, we can say Midjourney acts as a bridge between the creative minds at agencies and their clients' visions, facilitating a collaborative and dynamic creative process. It's more than just a tool for image generation; it can be a partner in the creative workflow, enhancing the ideation and moodboard creation process to meet the diverse needs of clients in the design and marketing world.

We now transition to a different yet equally fascinating use of this tool. We will explore how Midjourney brings a new dimension to storytelling, transforming written narratives into visually captivating experiences.

The visual power of storytelling

Midjourney has transformed the storytelling world by enabling us to craft immersive narratives that captivate audiences with outstanding visuals, bringing imaginative worlds and characters to life.

By using detailed prompts, Midjourney breathes life into stories, transforming them from linguistic experiences into interactive visual journeys. It seamlessly generates images that align with the narrative's essence, elevating storytelling and engaging readers and viewers on a deeper level.

Whether you are crafting novels, designing screenplays, or developing interactive narratives, Midjourney serves as a versatile ally, ensuring that imagery seamlessly integrates with your narrative. Its ability to comprehend complex prompts ensures a cohesive experience, immersing audiences in the story's world.

Imagine the chilling impact of a suspenseful horror story enhanced by Midjourney's spine-chilling imagery or the heartwarming camaraderie of a friendship story brought to life by vivid scenes of adventure. Midjourney seamlessly adapts to diverse genres, from gritty cyberpunk thrillers to whimsical children's tales. You just have to find the right words to describe what's on your mind.

In the end, it's fair to say that with Midjourney, storytelling potential expands exponentially, with narratives coming to life with unparalleled clarity and impact.

Let's imagine we are writing a fantasy story about a little girl's adventures with mythical creatures and magical landscapes. The first step is to define the visual style and consistency for all your images. This involves doing the following:

- **Creating a character sheet**: One effective approach in visual storytelling is to develop a character sheet for your protagonist. This sheet should showcase the character in different poses and experiencing different emotions, providing a visual reference that helps maintain consistency in character portrayal throughout the story. Always try to use clear and specific words to describe your character's looks and surroundings as this gives you more control and fewer options for Midjourney to explore by itself.

    However, it's important to note that even with these techniques, slight variations between the characters and scenarios generated in Midjourney are inevitable. This is due to the inherent randomness in Midjourney's output, a factor that we cannot fully control. While these variations can add richness and diversity to your narrative, they might also pose a challenge if you're seeking exact consistency in character appearance.

For those looking to achieve greater facial consistency for characters, I recommend exploring the advanced techniques we will discuss in *Chapter 9*. There, we explore different methods to fine-tune character appearances and maintain a higher level of consistency across different images generated by Midjourney.

- **Storyboarding key scenes**: Use Midjourney to storyboard key scenes, providing a visual narrative flow. This helps in visualizing the story's progression and key moments.
- **Utilizing reference images**: Incorporate reference images for composition and character consistency. This ensures that each frame aligns with the overall visual style of the story. Go to *Chapter 5* in case you need to remember how to use one or more images with a prompt.
- **Experimenting with styles**: Explore different artistic styles and parameters to find the one that best suits your story's mood and genre. After choosing the elements you want, always use them at the end of the prompts or consider adding them as a *prefer suffix*. This approach enhances the process of writing prompts by automatically appending these elements to your future prompts. For a detailed explanation of how to set up and use *prefer suffix*, for efficiency, refer to *Chapter 7*. Here is an example of what you can add as your *prefer suffix*: `in the style of whimsical fantasy, --ar 3:4 --s 250 --c 30 --seed 123abc`).

Let's see how we can practically apply the steps outlined here.

Practical example

Here is a story concept: A fantasy adventure about a little girl exploring a magical forest.

Character sheet prompt: **/imagine prompt** `character design sheet of a young girl with magical powers, various emotions and poses, in a fantasy forest setting.`

After choosing the image you feel better represents your character and story, what I like to do is click on the **Vary (Strong)** button to generate more ideas for the selected image. After that, I choose a second one I like to make a pair out of the two.

The visual power of storytelling   225

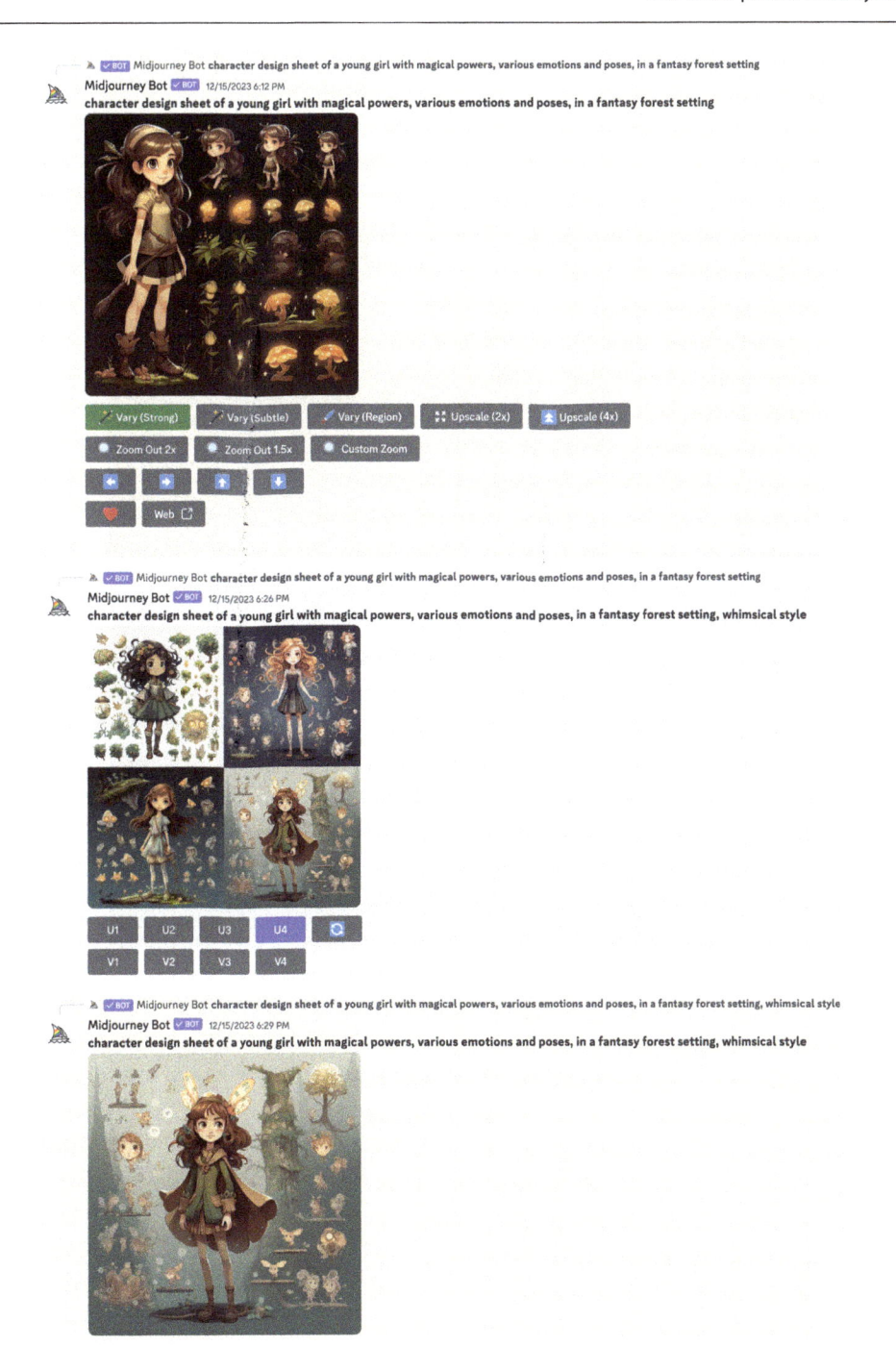

Figure 8.7 – The images that will be the base of my character

Now what I do with my two generated images is use them as part of an image prompt. I copy the URLs of the images and paste them followed by my prompt:

/imagine prompt `<image url><image url>young girl with magical powers, encountering mythical creatures in a fantasy forest, whimsical style:`

Figure 8.8 – Image generated using /imagine prompt - https://s.mj.run/Ktg6Rolsudk https://s.mj.run/xwLW00ZgiGw young girl with magical powers, encountering mythical creatures in a fantasy forest setting, whimsical style

After establishing the initial prompt, I engage in an iterative process of refinement using Midjourney's features. This involves frequently using the **Vary (Subtle)** button to introduce minor variations to each prompt iteration. Simultaneously, I keep the *Remix* option active, which allows me to make direct edits to my prompt for each variation. This combination of subtle variation and active editing enables me to fine-tune the visuals, ensuring they align closely with my envisioned narrative.

Using the **Vary (Subtle)** button and editing the prompt to **/imagine prompt** `<image url><image url> young girl with magical powers, playing with a cute dragon in a fantasy forest setting, whimsical` yields the following:

Figure 8.9 – This is the output of my prompt, combining an image url with text and seed /imagine prompt - https://s.mj.run/Ktg6Rolsudk https://s.mj.run/xwLW00ZGiGw young girl with magical powers, playing with a cute dragon in a fantasy forest setting, whimsical --seed 495881614

Continuing the process of refining my visuals, I again utilize the **Vary (Subtle)** button in conjunction with the active *Remix* feature. This time, I edit the prompt to be **/imagine prompt** `<image url><image url> path to a wooden cabin in a fantasy forest setting, whimsical --seed 495881614`.

After pressing *enter* and generating the images, I select my favorite one.

In my chosen image, the girl still appears, but my focus is on depicting the path to the house. To adjust this, I use the **Vary (Region)** button. I select the part of the image I want to change, which in this case is the area excluding the girl, to emphasize the path to the cabin. I leave the message box empty and hit *enter*:

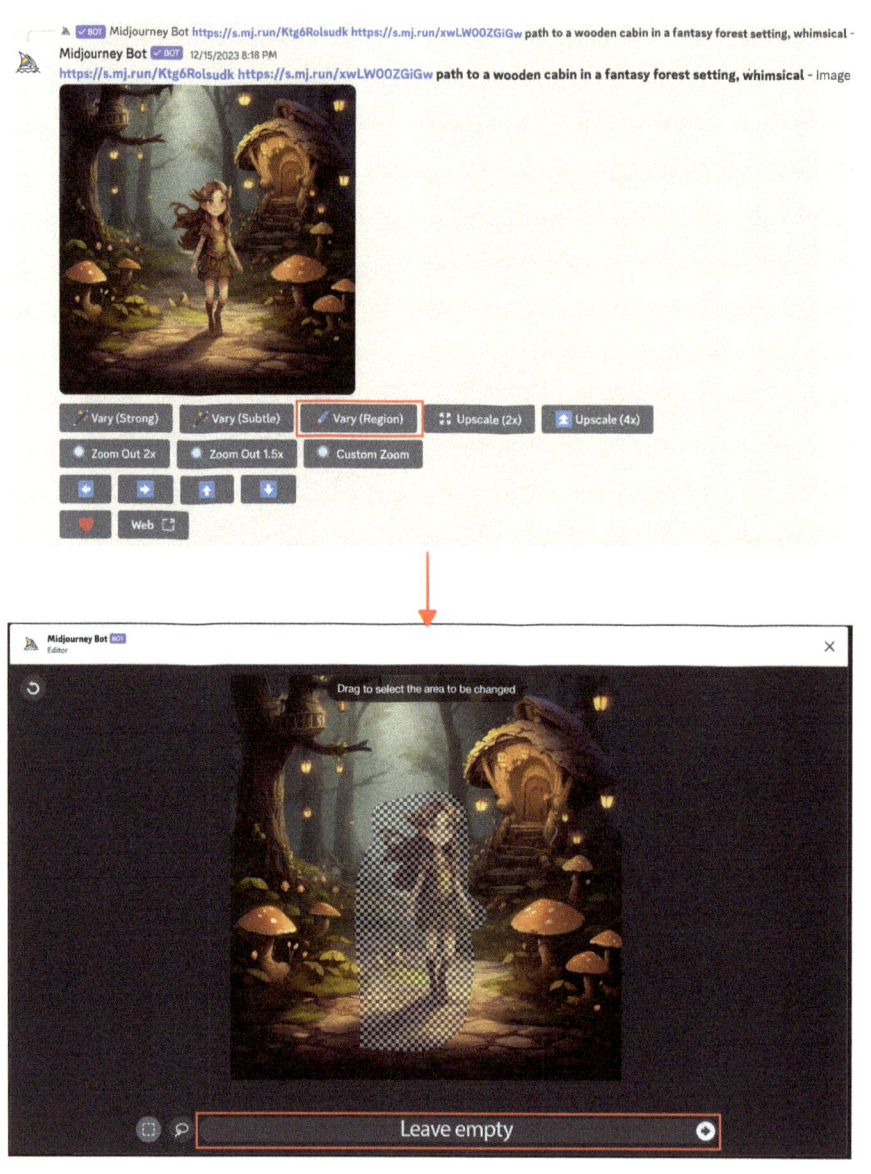

Figure 8.10 – Using the Vary (Region) option to modify my image

After pressing *enter*, Midjourney generates a new set of images based on this specific region selection, giving me the exact visual focus I need:

Figure 8.11 – The end result is the same image but now without the character

By following these steps and utilizing Midjourney's diverse features, you can create visually striking narratives that deeply resonate with your audience. An additional tip for achieving consistency in your visual storytelling is to use the `--seed` parameter (if you cannot recall how to use this parameter, go back to *Chapter 4* to learn how to use the `--seed` parameter). By specifying the *seed* of an image you prefer, you can guide Midjourney to produce variations with a similar aesthetic, maintaining a consistent look and feel across your narrative. You also may consider using image weight, the `--iw` parameter, (check *Chapter 4* to revisit the parameter list), in the prompt to help emphasize the image as a reference.

Furthermore, consider leveraging the Style Tuner to create a unique style that can be applied to all your images. This not only enhances consistency but also adds a distinct visual signature to your storytelling. Once you've crafted a style with the Style Tuner, you can apply this customized style to your entire narrative, ensuring that each image reflects a cohesive artistic vision.

Next, we'll shift our focus to the art of creating icons and logos. This section will guide you through the process of designing visually compelling brand sets with Midjourney, helping you craft icons and logos that effectively embody brand identities and communicate their core messages.

Creating brand sets with Midjourney – icons and logos

Midjourney is not only a powerful tool for creating visually stunning images and narratives; it also excels in designing captivating icons and logos that effectively embody brand identities and communicate core messages. In the following subsections, we will explore the various aspects of this process.

We'll start by exploring the fundamental concepts of brand identity, including how to identify a brand's core values, target audience, and aesthetic direction. This section will focus on crafting effective prompts in Midjourney to create icons and logos that resonate with your brand's essence. We'll also examine the Midjourney features that enable precise control and refinement of your designs, ensuring they align seamlessly with your brand's identity. Additionally, we will learn how to create different types of icons that complement and enhance your brand.

By covering these topics, we aim to provide a comprehensive guide on using Midjourney to create impactful and cohesive brand sets, including icons and logos.

Understanding the essence of brand identity

Before creating icons and logos, it's crucial to understand the brand's core values, target audience, and overall aesthetic direction. This foundational knowledge guides the design process, ensuring that each element aligns with the brand's identity.

Effective prompt crafting is essential in translating a brand's essence into visual symbols. Midjourney requires clear, concise prompts to generate icons and logos that capture the brand's spirit. For instance, a prompt for an eco-friendly brand might include keywords such as *sustainable*, *green*, and *organic* combined with visual descriptors. To illustrate this process, I will conduct an experiment to create some logo ideas for a vegan restaurant. I'll start with three different prompt approaches, each designed to explore various aspects of the restaurant's identity and values. This experiment will demonstrate how Midjourney can be used to generate a range of creative concepts, providing a starting point for developing a compelling and representative logo for the brand.

In my first prompt, I will utilize Midjourney to generate a set of logos. At this stage, I'm seeking ideas and inspiration, so I'll be using Midjourney as my creative partner for a brainstorming session focused on vegan logos:

/**imagine prompt** a collection of minimalist vector logos, vegan food restaurant, vegan brand, simple shapes and lines, pure white isolated background, organic food, vegan restaurant, vegetarian, business logo

Let's see what we get:

Figure 8.12 – My first prompt approach

Now that I've seen some examples with a specific prompt, I'm curious to see how Midjourney responds to a more open-ended request. In this next prompt, I'll provide less information about styles and other specifics. This approach will test Midjourney's creative range when given broader, more vague directions:

/imagine prompt `vector set of logos for restaurant's vegetarian food`

Let's see what Midjourney generates with this different type of prompt:

Figure 8.13 – My second prompt approach

This last prompt has provided valuable insights, particularly highlighting the need for specific colors and elements in the logo.

While the restaurant is vegan, I realize that opting for an all-green color scheme might be too obvious. Now that we have a starting point, it's time to enhance our prompt with more details. This next step involves incorporating specific design elements and color choices that align with the restaurant's identity, yet steer clear of clichéd representations:

/imagine prompt `vegan restaurant logo with a leafy floral background, in the style of hand-drawn elements, simple, colorful illustrations, elaborate fruit arrangements, soft color palettes, flat composition, minimal graphic vector logo for a vegan food plant-based restaurant, flat vector art, logo vector, outlines, plain white background:`

Figure 8.14 – My third prompt approach

Now that we've added precise details to our prompt and are almost satisfied with the results, it's time to address the unexpected words and letters appearing at the center of the logo.

Leveraging Midjourney's features for precise control

As we've seen before, the **Vary (Region)** button is a useful tool for refining our images. By selecting the area with the unwanted text and typing `plain flat background color` in the message box, we can effectively remove these elements:

Figure 8.15 – How to easily erase unwanted elements

Additionally, the `--no` parameter can be a handy tool to prevent specific elements such as text from appearing in the first place. For example, using `--no newspaper text` in your prompts can help avoid such issues. However, it's important to note that even with the `--no` parameter, unwanted results can sometimes occur. This is where Midjourney's flexibility becomes invaluable. The ability to make adjustments within Midjourney, such as using the **Vary (Region)** feature, is incredibly useful for fine-tuning your design to perfection.

Unfortunately, at the time of writing this, Midjourney does not yet handle typography perfectly. For projects such as logos where text is a crucial element, this can be a limitation. In my case, to incorporate the brand name and to fine-tune the logo, I turned to Adobe Illustrator, a tool I'm comfortable with for typographic and logo-making work. However, you can use any other software that suits your needs for this part of the design process.

Here is the final result for my brand logo, a blend of Midjourney's creative AI capabilities and the precision of traditional graphic design software:

Figure 8.16 – My final logo after refining and perfecting it in Illustrator after the initial conceptualization with Midjourney

Now that we have the foundation of our logo, it's time to extend our brand identity by creating matching icons. Icons are distinct from logos in that they are small, symbolic images used to represent concepts, actions, or objects within a brand's visual language. While logos often serve as the centerpiece of a brand's identity, icons are used to complement and reinforce the brand's message across various applications, such as in digital interfaces, marketing materials, and product packaging.

To achieve a cohesive look between our logo and icons, we'll employ a combination of images, text prompts, and the *seed* parameter in Midjourney. This approach ensures that our icons not only match the style and essence of our logo but also enhance the overall visual storytelling of our brand.

First, react to your logo image with the envelope emoji to retrieve the seed. This seed is crucial as it helps maintain stylistic consistency in the icons we're about to create. Next, copy the image URL of your logo that was generated by Midjourney.

Creating brand sets with Midjourney – icons and logos | 235

With the image URL and seed in hand, we're ready to craft our prompt. This prompt will be tailored to generate icons that not only align with our brand's aesthetic but also complement our newly designed logo. Here's how we adapt our prompt to create our brand iconography:

/imagine prompt `[image url] [insert element you want to be represented by the icon] hand-drawn icon --seed [value]`

In my case, the goal is to develop a series of icons that can be used across different mediums, enhancing the brand's presence on various platforms. These icons will be versatile enough for social media, our website, menus, and even for visually representing different dishes. This diversity ensures that our brand identity is consistently represented, no matter where it appears.

To create this diverse set, I'll tailor my prompts in Midjourney to generate icons that are not only visually aligned with our brand but also functional for each specific application. Whether it's an icon for Instagram, a graphic for our website's home page, or a symbol to represent a signature dish, each icon will be thoughtfully designed to serve its purpose effectively while maintaining the aesthetic coherence of our brand.

I will start by creating a set of icons to illustrate my dessert menu offers:

/imagine prompt `https://s.mj.run/jkb4YHAobEQ dessert icon set, hand-drawn icon --seed 1999839074:`

Figure 8.17 – Icon set example for desserts

236 Exploring Practical Use Cases and Pushing Boundaries

Let's create a similar set for our beverages menu:

/imagine prompt `https://s.mj.run/jkb4YHAobEQ beverage icon set, hand-drawn icon --seed 1999839074s:`

Figure 8.18 – Icon set example for drinks and beverages

Now, I'm intrigued to explore more. I'm keen to see the results when I request a more comprehensive set from Midjourney, one that encompasses all the diverse concepts featured in my restaurant's menu. This will not only test Midjourney's versatility but also provide a fascinating glimpse into how it interprets and visually represents the full range of our menu's offerings:

/imagine prompt `https://s.mj.run/jkb4YHAobEQ dessert, fruits, vegetables, drinks icon set, hand-drawn icon --seed 1999839074:`

Figure 8.19 – Icon set example for menus

There are always a myriad of options for creating iconography with different artistic approaches in Midjourney. Ultimately, the style you choose should align with your brand's identity and the message you wish to convey. Here are some examples of icons created with Midjourney, each showcasing a different aesthetic:

- **Modern linear icons**: **/imagine prompt** `blue house, modern line icon, bold outline, solid color, pixel perfect, pure white background`:

Figure 8.20 – A simple house icon example, demonstrating a modern and clean linear style

- **Architectural style**: **/imagine prompt** `a house design in a thin line, in the style of bold outlines, flat colors, naive style, minimalist illustrator, light gray and bronze, light yellow and emerald, 35mm lens, meticulous design, white background`:

Figure 8.21 – An icon example in a specific style, reflecting an architectural and detailed approach

- **Mobile app icon**: **/imagine prompt** `orange clock icon, square with round edges, flat design icon, minimalist, pure white background`:

Figure 8.22 – An icon designed for mobile apps, showcasing a minimalist and functional design suitable for app interfaces

These examples illustrate the flexibility of Midjourney in creating icons that range from simple and modern to detailed and stylistic. By adjusting your prompts, you can explore a wide range of aesthetics and find the perfect match for your brand's visual language.

Having explored the diverse possibilities of icon and logo creation with Midjourney, let's now shift our focus to the next crucial aspect of branding and marketing: **product mockups**. In the following section, we'll examine how Midjourney can be effectively used to create realistic and engaging mockups, showcasing our products in a visually striking way.

Figure 8.23 – Bonus image /imagine prompt - Kawaii vibrant 3D food icons

Utilizing Midjourney's photorealism for product mockups

Now that we have developed a cohesive set of logos and icons for our brand, the next step is to integrate these elements into product mockups. Midjourney's ability to generate photorealistic images opens up exciting possibilities for creating stunning visuals that closely resemble real-life photography. This approach not only complements our brand's visual identity but also enhances our overall marketing strategy with the following:

- **Enhanced realism**: Lifelike images provide a more authentic representation of products, allowing potential customers to visualize them in great detail.

    /imagine prompt `Modern minimal design for trendy wellness brand, body and mind, luxury brand, modern package design, muted colors, Product photography, retail display:`

    Figure 8.24 – Professional and modern product display prompt /imagine prompt - Modern minimal design for trendy wellness brand, body and mind, luxury brand, modern package design, muted colors, Product photography, retail display

- **Brand identity reinforcement**: Realistic and high-quality photography reinforces brand identity by showcasing products in their natural settings, aligning with the brand's visual style and aesthetics. This can effectively convey the product's story, transforming standard mockups into captivating visual narratives.

```
/imagine prompt a ripe avocado is sitting on a natural wooden
placemat, in the style of sculpture-based photography, Japanese
minimalism, nature-inspired installations, backlit photography,
juxtaposition of objects, vray tracing, organic stone carvings:
```

Figure 8.25 – Realistic food display prompt using /imagine prompt - a ripe avocado is sitting on a natural wooden placemat, in the style of sculpture-based photography, Japanese minimalism, nature-inspired installations, backlit photography, juxtaposition of objects, vray tracing, organic stone carvings

Midjourney's latest versions offer robust and photorealistic image creation capabilities. Detailed scene descriptors and specific lighting conditions can be used to capture the essence of your product. Descriptors in your prompt such as *wildlife portrait*, *time-lapse*, or *high-speed photograph* can be combined with lighting conditions such as *natural light*, *golden hour*, or *studio lights* to create the desired atmosphere. Incorporating specific camera types, film styles, and camera settings into your prompts can also enhance the effect.

Having explored how Midjourney enhances brand identity through realistic imagery, let's now turn our attention to its specific applications in food photography.

Practical applications

This section will illustrate how Midjourney's features can be employed to create a varied collection of images, each reflecting different facets of our vegan restaurant's offerings. Here are some food photography prompts, adaptable to various subjects:

- **Top-view food photography**: **/imagine prompt** `top-view food photography of freshly grilled broccoli on a wooden plate in a bohemian and modern vegan kitchen, dreamlike lighting, pastel green background, high detailed, photo detailed, Professional food photography:`

Figure 8.26 – Food prompt example

- **Product photography:** **/imagine prompt** `delicious beetroot burger with lettuce on a white background, in the style of studio food photography:`

Figure 8.27 – Burger prompt example

- **Knolling photography:** **/imagine prompt** `knolling photography of gourmet spices and ingredients, on a rustic wooden table:`

Figure 8.28 – Knolling food prompt example

/**imagine prompt** knolling photography of fruits and vegetables over a clean light background

or

/**imagine prompt** Diagram of fruits and vegetables over a clean light background:

Figure 8.29 – Diagram food prompt example

- **Editorial photography:** **/imagine prompt** `vegan toppings in containers (onion, mushroom, tomato, orange), Cinematic, Editorial Photography, Photography studio light:`

Figure 8.30 – Editorial photography food prompt example

- **Interior photography**: **/imagine prompt** `interior photo of bohemian and fashion vegan food restaurant, lunch time:`

Figure 8.31 – Interior design prompt example

Each of these categories represents a unique approach to visualizing our restaurant's culinary delights and ambiance, showcasing the power of Midjourney in creating compelling visual narratives for any related businesses.

Creating professional product mockups with Midjourney

To craft professional product mockups, begin with a basic concept and enhance it by incorporating specific details, settings, and contexts. Midjourney's text-to-image prompt feature is instrumental in producing highly detailed visuals that effectively showcase your products.

For example, to create a mockup for a hand-held product against a specific background, use a prompt such as `hand holding, [Product] mockup, [Color] backgrounds, handheld, [aerial, front, bottom, etc.] view, product display.`

Let's create an illustration that matches the logo we generated for the vegan restaurant, adapting it for various mockups automatically created in Midjourney. We'll also use the `--tile` parameter, as discussed in *Chapter 4*.

Start with the following prompt: **/imagine prompt** `vegetables and fruits hand-drawn illustrations, simple, soft color palettes, elaborate fruit arrangements --tile`.

After selecting and upscaling the preferred image, we can further customize it. For instance, using **Custom Zoom**, we can edit the prompt as follows: **/imagine prompt** `Framed in a wall art sitting in the middle of bohemian and fashion vegan food restaurant --ar 1:1 --zoom 2`:

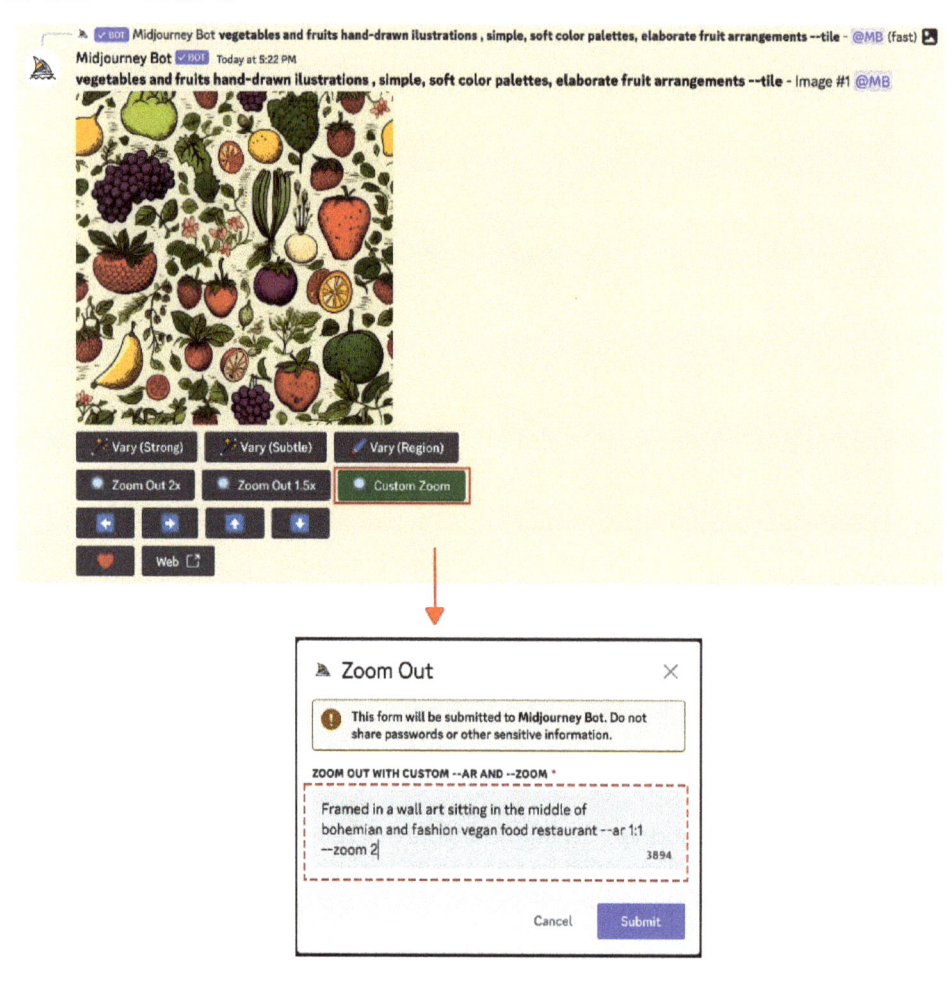

Figure 8.32 – Interior mockup

This quickly starts to shape our space with a cool brand aesthetic:

Figure 8.33 – My image framed on the restaurant wall

Let's explore more mockup adaptations for different services and purposes.

In one instance, I used the `printed on a tote bag, displayed in Photo vertical green wall --ar 1:1 --zoom 2` prompt:

Figure 8.34 – Custom-designed illustration printed on a Tote Bag

248　Exploring Practical Use Cases and Pushing Boundaries

I also tried it with the `printed on a wide computer screen, on a concrete surface, with plants beside --ar 1:1 --zoom 2` prompt:

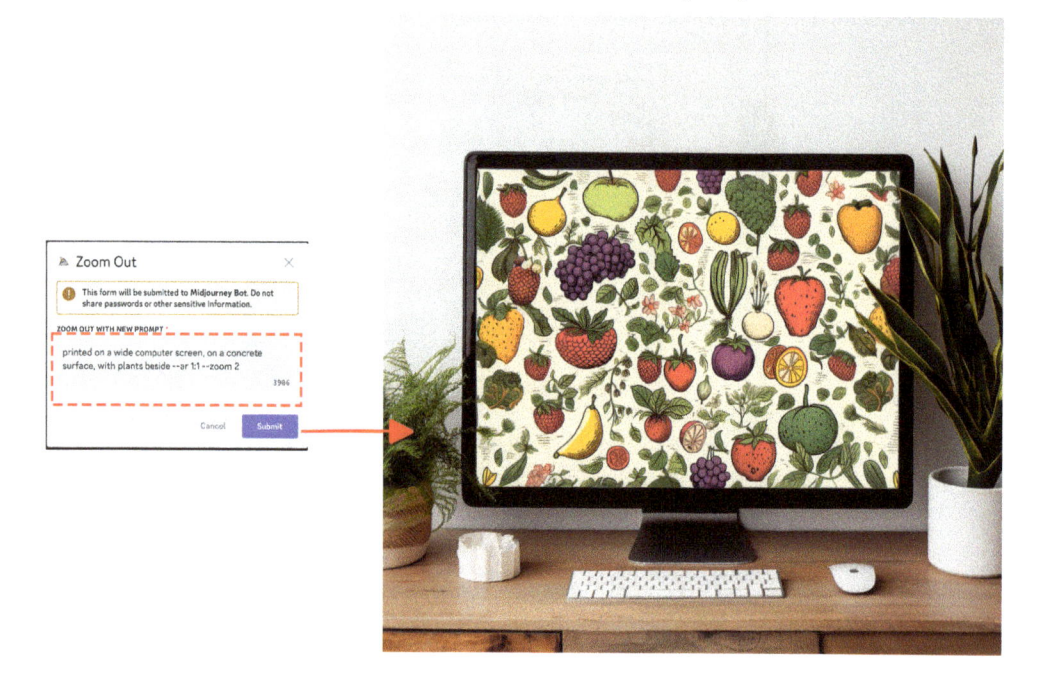

Figure 8.35 – Adapting my image to any medium

Another approach to creating versatile mockups in Midjourney involves using text prompts that create empty spaces in the image. These spaces can be used for advertising purposes or to insert white labels on products, which are easily editable with third-party tools.

Here are just a few examples of how to make use of these white spaces:

- **/imagine prompt** `Mockup empty, blank book, a hand holding the phone, busy street background`:

Figure 8.36 – iPhone mockup

- **/imagine prompt** `Mockup empty, blank iphone screen, on a clean background:`

Figure 8.37 – iPhone product mockup

- **/imagine prompt** `Mockup empty, blank label design for modern trendy cosmetics in white simple bottles, on a clean white background, with vase of flowers behind:`

Figure 8.38 – Product mockup

- **/imagine prompt** `Mockup empty, a blank white billboard in the middle of the highway, high-speed Photography of city traffic lights:`

Figure 8.39 – Advertising mockup

- **/imagine prompt** `Mockup empty, custom T-shirt mockup, in the style of soft tones, high resolution, apparel mockup:`

Figure 8.40 – Apparel mockup

- **/imagine prompt** `Mockup empty, blank business card, sitting on a concrete table, Scandinavian clean ambience, daylight and clear shadow:`

Figure 8.41 – Business card mockup

By making use of these Midjourney capabilities, we can create a wide range of mockups that are easily adaptable to our brand's needs, enhancing our marketing and visual presentation.

Summary

In this chapter, we explored the diverse applications of Midjourney in branding and marketing. We tested its use for creating moodboards and storytelling, enhancing the visual narrative of brands. The focus then shifted to designing cohesive brand sets, including icons and logos, leveraging Midjourney's capabilities for photorealistic product mockups.

We examined Midjourney's practical applications in various real-world scenarios, highlighting its revolutionary role in diverse industries from graphic design to product development. The chapter emphasized the importance of maintaining consistency across various mediums, showcasing Midjourney's ability to generate cohesive visual elements. We can say that this chapter marks a significant milestone in our journey with Midjourney, equipping us with the knowledge and skills to fully use its potential in our creative projects.

As we conclude, this chapter has equipped us with essential skills to utilize Midjourney effectively, setting the stage for further discovery in the next chapter.

9
Unlocking Tips and Tricks and Special Functions

As we venture further into Midjourney's capabilities, we'll uncover a hidden cache of techniques and functionalities that can transform your creative work. This chapter serves as your guide to unearthing special features designed to elevate your Midjourney experience. Learn how to effectively utilize faces, master the art of generating high-resolution images suitable for professional printing, and decipher ultimate tricks on the Midjourney art of prompting.

In this chapter, we will cover the following topics:

- **Using the face you want**: Here, we will discover how to consistently generate specific faces in your creations, an essential skill for character consistency in storytelling and branding
- **Creating high-resolution imagery for printing**: In this section, we will learn about the latest updates made available for resolutions in Midjourney and explore techniques for upscaling images to meet professional printing standards
- **Tips and tricks to improve your prompts**: A collection of tips and tricks will be unveiled here to refine your understanding and control of Midjourney's capabilities

By the end of this chapter, you will have a deeper understanding and enhanced control of Midjourney, enabling you to apply the taught techniques to your creative projects. Get ready to discover the world of specialized functions and tips that will transform your Midjourney experience!

Technical requirements

To make the most of this chapter and follow along with the hands-on activities, you'll need to ensure you've set up everything we covered in *Chapter 2*. Here's a quick recap:

- **Midjourney account**: You'll need an active account on Midjourney. If you haven't registered yet, refer back to *Chapter 2* for a step-by-step guide.
- **Discord installation**: Ensure you have Discord installed on your device, as we'll be working with it alongside Midjourney. Again, the installation guide is available in *Chapter 2*.
- **Payment plans**: At the moment, Midjourney has no free or trial plan. A Midjourney paid plan is necessary. Please ensure you've selected the most suitable plan for your needs.
- **Server creation**: Having your own server can significantly enhance and expedite your experience with Midjourney. If you're yet to set up a server in Discord, *Chapter 2* covers the process in detail.

Using the face you want with InsightFaceSwap Bot

The ability to consistently generate characters with specific facial attributes is a valuable skill for any Midjourney user, regardless of their experience level. This is especially true for those who want to create characters that are consistent with their vision or that represent a particular character from a story or other creative work.

So, let's explore a technique for controlling facial features in Midjourney through the use of a third-party tool, **InsightFaceSwap Bot**. This tool, based on the open source InsightFace library (`https://insightface.ai/`), offers advanced face analysis capabilities for both 2D and 3D images. InsightFaceSwap Bot's true power lies in precision face editing and seamless character continuity, ensuring your character looks recognizable and true to form in every scene. It's particularly useful to maintain facial consistency across different images, enhancing character continuity in your projects.

Let's follow this step-by-step guide to face swapping with InsightFaceSwap Bot in Midjourney:

1. Let's begin by adding the InsightFace Bot to your Discord server. This is an essential step for initiating the face-swapping process. You can invite the bot by navigating to `https://discord.com/oauth2/authorize?client_id=1090660574196674713&permissions=274877945856&scope=bot` and selecting your desired server; in my case, it will be `myImagineServer`. Then, click **Continue**:

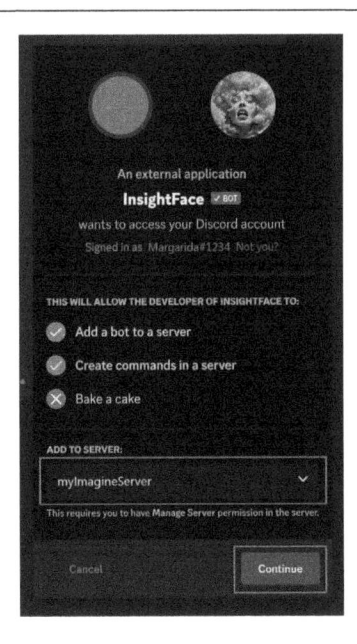

Using the face you want with InsightFaceSwap Bot   255

Figure 9.1 – Inviting InsightFace Bot to your server

2. Confirm the necessary permissions for the InsightFace Bot on your server by clicking on **Authorize**:

Figure 9.2 – Authorizing InsightFace Bot to access your server

3. We then complete the CAPTCHA to verify that we are not a robot to grant the bot access to your server:

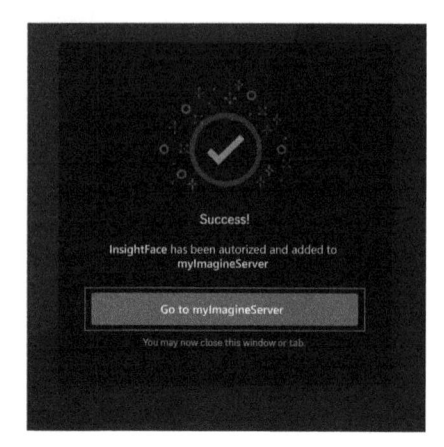

Figure 9.3 – After confirming you are not a robot, the bot has access to your server

After adding the InsightFace Bot to your server, let's start by registering our character identity using the /**saveid** command. This identity is then used for all your future face-swappings. Remember to use clear, front-view photos for better results.

4. Start by writing /**saveid** in the chat box and hitting *enter*. Then upload your image and choose an **idname** for it:

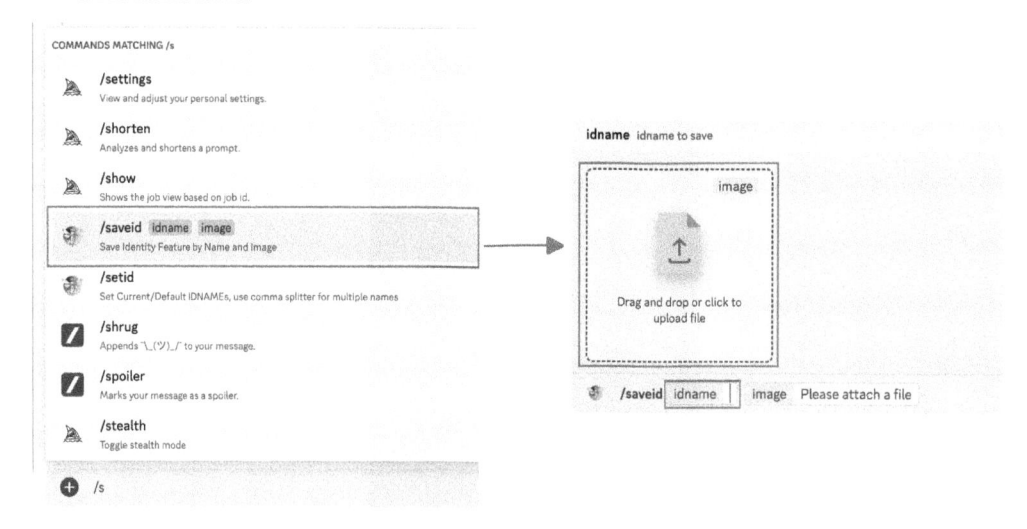

Figure 9.4 – Using the /saveid command to start and register your character

5. I'm going to upload an image of a character named `Peter` (the idname can be up to 10 characters long, consisting of letters or numbers) that I previously generated on Midjourney, and I want to use as a character for a Peter Pan story I'm developing:

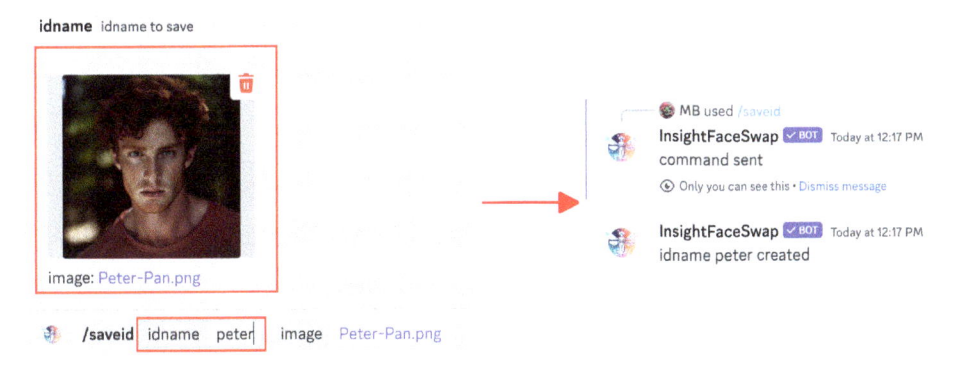

Figure 9.5 – Uploading the image you want your character to resemble and registering the identity

6. After uploading the image and assigning an idname, press *enter*. You will receive two confirmation messages indicating that the command was sent and the idname was created. This registered identity will be used for all future faceswaps.

   Now, let's start using the **INSwapper** feature!

7. Start by generating an image in Midjourney. Once you have a grid of images, pick your favorite and upscale it using the **U** button below the image as you would normally do. In my case, I used the following prompt:

   /imagine prompt `a futuristic red hair Peter Pan, blending classic elements with cyberpunk aesthetics. neon-lit skylines, high-tech gadgets replacing traditional fairy dust, and sleek, hoverboard-like flying devices. His outfit is a fusion of traditional green with smart fabrics displaying shifting patterns. The Lost Boys are now a diverse group of tech-savvy youths, skilled in virtual reality and urban exploration in a sprawling, vertical city.`

258 Unlocking Tips and Tricks and Special Functions

8. Now, right-click on the upscaled image, and from the **Context** menu, select **Apps** and then choose **INSwapper**:

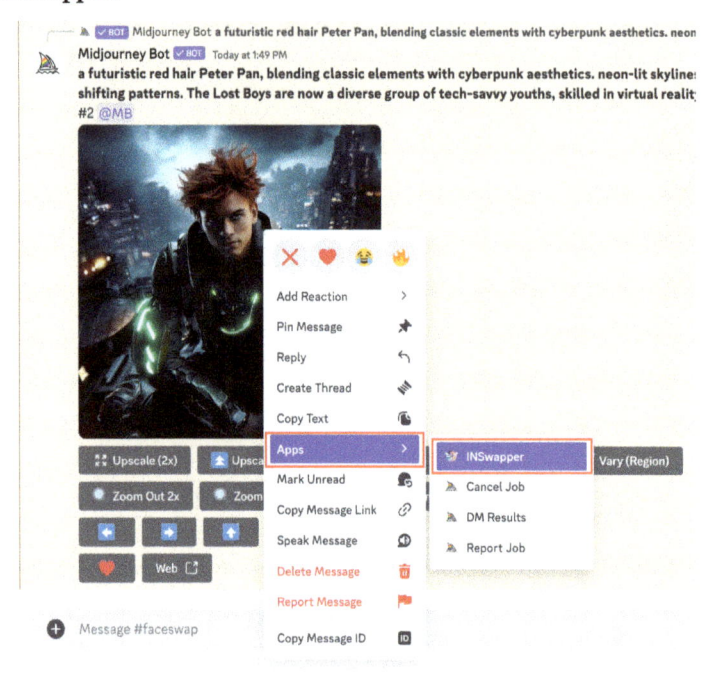

Figure 9.6 – Selecting Apps and INSwapper from the menu

After selecting **INSwapper**, the process will begin automatically. You'll receive a notification confirming that your command has been sent. Shortly after, the transformed image with the swapped face will appear below the notification:

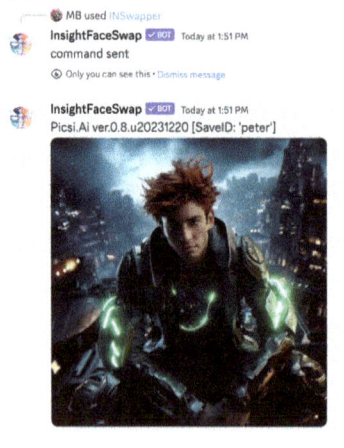

Figure 9.7 – The result will appear almost instantly

This simple process allows you to seamlessly integrate the chosen face into your Midjourney-generated image, offering a quick and effective way to personalize your creations. In *Figure 9.8*, you can see some of the images I generated using this feature:

Figure 9.8 – Examples of face consistency across different images

Now, I repeated the process I used to create the `Peter` character for a character named `Wendy`. Although INSwapper is a paid tool, it offers a free plan that includes 50 credits per day. This means you can swap up to 50 faces daily and create up to 20 face idnames, with one morph face allowed per image:

Figure 9.9 – Using the /saveid command again and pressing Enter to create another id image

After adding `Wendy` as a new idname to my list, I used the same prompt and seed as I did for `Peter` to ensure a similar style. I made slight adjustments to the prompt to also introduce some pink colors and create a feminine character. Once I'd selected my favorite images from the new grid and upscaled them, I used the INSwapper feature just as before:

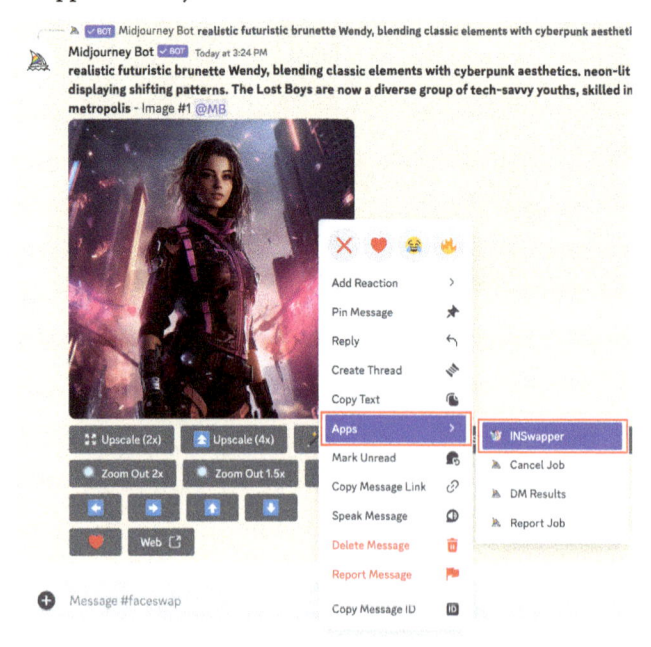

Figure 9.10 – Selecting Apps and then INSwapper from the menu

By right-clicking on the upscaled image, selecting **Apps**, and then choosing **INSwapper**, I successfully created consistent faces for the two main characters of my story:

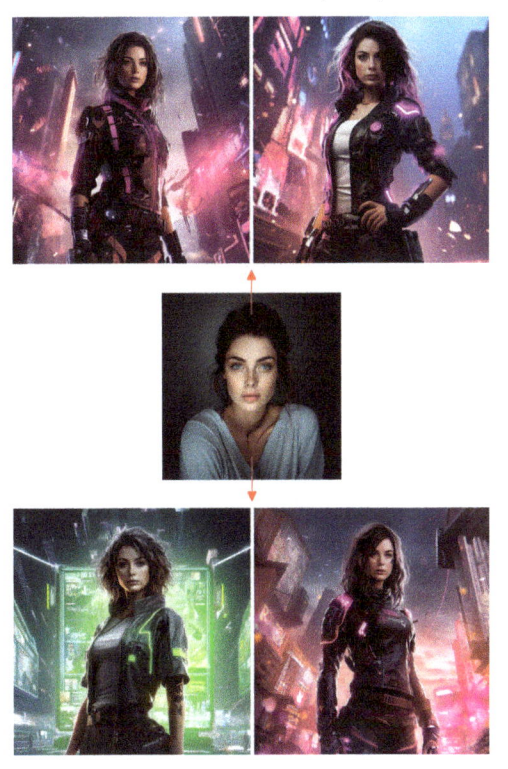

Figure 9.11 – Illustrations of uniform facial features in varied images

To help you easily understand some of the commands that this tool also offers, here is a simple list of the key commands of InsightFace:

- **/saveid** `name upload-id-image`: Upload and register an id image (preferred id photos: front view, high quality, without glasses or heavy bangs).
- **/setid** `name(s)`: Set default identity name(s) for image generation. Every time you need to change the character, use this command followed by the idname you want to work with.
- **/listid**: List all registered identities names (up to 20), the current idname in use, the type of subscription you have, and how many credits you have left.
- **/delid** `name`: Delete a specific identity.
- **/delall**: Delete all registered identities.

For access to additional features offered by this app, consider subscribing to one of their paid plans. More detailed information about the subscription options is available on their Patreon page (`https://www.patreon.com/picsi`) where you can select the plan that best suits your needs.

So far, we've explored how to control facial features in Midjourney, enabling you to create more realistic and consistent characters. However, if you want to bring these creations to life in high-resolution prints, you'll need to consider upscalers and external applications. That's where we'll head next.

Creating high-resolution imagery for printing

Before we dive into the new Midjourney model version V5.2 and model version V6 [ALPHA] upscalers and third-party tools to enhance image quality, it's essential to grasp the fundamental concept of image resolution. Understanding this concept will enable us to utilize Midjourney's capabilities effectively, particularly when it comes to producing high-quality images for various applications:

- **What is image resolution?** Image resolution refers to the level of detail an image contains, often depicted by the number of pixels per unit area. Imagine an image as a grid composed of tiny squares, with each square representing a pixel. The greater the number of pixels in this grid, the more refined and clear the image will be.

- **Are file size and image dimensions the same?** No, the file size and image dimensions of an image are not the same. The file size of an image is determined by the amount of data it contains, which is influenced by its resolution. Higher-resolution images contain more pixels, which typically results in more data and a larger file size. On the other hand, image dimensions refer to the number of pixels in the width and height of the image, expressed as width x height (e.g., 1024 x 1024 pixels). For example, a 4K image has dimensions of 3840 x 2160 pixels. It's important to note that file size is also affected by factors such as the image's format and compression.

- **What does DPI stand for?** DPI stands for **dots per inch**. It measures the density of pixels (dots) packed within a physical space, typically an inch. DPI is a crucial factor in printing, where a higher DPI means that more dots are packed into each inch, resulting in greater detail and sharper images. However, for digital displays, the DPI is less relevant; what matters more are the pixel dimensions of the image (i.e., the width and height measured in pixels).

- **Are there any new upscalers in Midjourney?** Absolutely! As I am writing this book, it's thrilling to see the evolution of Midjourney's upscaling capabilities. For the current default model version 5 (including V5.1, V5.2, and the Niji model V5), the introduction of **Upscale 2x** and **Upscale 4x** has already provided unprecedented flexibility and quality in image enlargement. However, the innovation doesn't stop there. The recent rollout of V6 [ALPHA] introduces its own set of improved upscalers, complete with both **Subtle** and **Creative** modes, each designed to increase the resolution by two times.

- **What are the biggest Midjourney image sizes?** In Midjourney, the standard image size is typically 1024 x 1024 pixels. However, with the introduction of the new upscaling tools, you can enhance the resolution of these images. Upscaling expands the image size to higher resolutions such as 2048 x 2048 or 4096 x 4096 pixels, enabling larger prints without compromising image quality.

- **Is it possible to print high-quality Midjourney images?** Yes, it is possible to print high-quality images created by Midjourney. The key to achieving a high-quality print is to ensure that the resolution of your image is suitable for the desired print size. For example, an image with a resolution of 1024 x 1024 pixels can produce a sharp print of about 3.4 x 3.4 inches (approximately 8.6 cm x 8.6 cm) at 300 DPI. If you upscale the image to a resolution of 4096 x 4096 pixels, you can get a high-quality print of around 13.7 x 13.7 inches (about 34.3 cm x 34.3 cm) at the same DPI, but you can push these sizes much further. This upscaling ability makes Midjourney images versatile for both digital displays and print media.

Next, we'll explore these new upscalers in more detail and demonstrate how to effectively apply them to your Midjourney projects.

Introducing Midjourney's new upscalers – Upscale 2x and Upscale 4x

Midjourney has recently enhanced its capabilities with the introduction of two new upscaling features: Upscale 2x and Upscale 4x. These features are designed to significantly improve the resolution of your images, allowing for more detailed and larger prints without compromising on quality. The **2x** and **4x** upscale buttons are conveniently located under any normal image upscale from any V5, V5.1, V5.2, or Niji job (currently, these upscalers are not compatible with the *pan* feature or the `--tile` parameter):

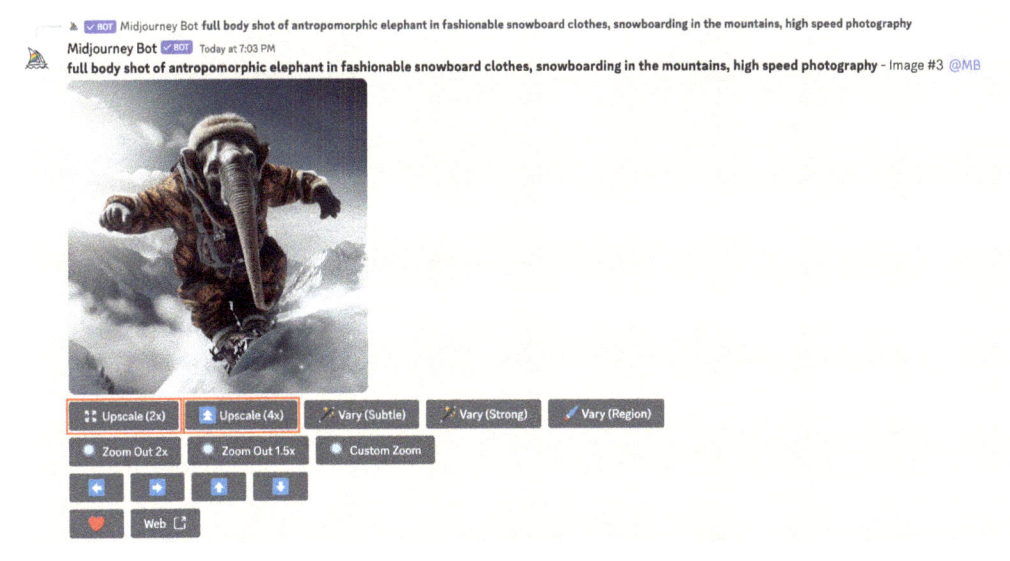

Figure 9.12 – Upscale any image from your grid in model version 5 or Niji and the two new buttons will appear

These upscalers are designed to preserve the original image's details as much as possible. However, it's important to note that while they excel in enhancing resolution, they may not always rectify existing glitches or issues in the original image.

The efficiency of these upscalers is measured in terms of GPU usage and time. Upscaling an image with Upscale 2x consumes about twice the GPU minutes and time compared to generating an initial image grid. In contrast, Upscale 4x, while offering a higher resolution, requires approximately six times the duration and three times the GPU minutes of Upscale 2x. This increased resource usage is a crucial factor to consider, especially since these upscalers currently operate with fast GPU time. Midjourney is actively exploring the feasibility of making these features available in *Relax* mode, which would be a more resource-efficient option.

Given these considerations, it's advisable to use the Upscale 2x and 4x features as the final steps in your image creation process. Before you decide to upscale, ensure that the final image meets your satisfaction, as the upscaling process will preclude further modifications using inpainting and outpainting features.

Figure 9.13 – From left to right you can see the difference between the normal upscaler (1024 x 1024 pixels), Upscale 2x (2048 x 2048 pixels), and Upscale 4x (4096 x 4096 pixels)

Another key aspect to remember is that the upscalers take into account the image prompts used during the initial creation. This means the quality of images included in the prompt can influence the overall quality of the upscaled output. Therefore, selecting high-quality prompts is essential for achieving the best possible results with these new upscaling features:

Figure 9.14 – From left to right you can see the level of definition between the normal upscaler, Upscale 2x, and Upscale 4x

Exploring new upscaling options in Midjourney V6 ALPHA

Although currently in the alpha stage, which means it is still under development and might exhibit occasional inconsistencies, **Midjourney V6 ALPHA** introduces refined upscaling options that broaden the scope of image enhancement. These options, **Upscale (Subtle)** and **Upscale (Creative)**, offer distinct approaches to improving image resolution. And like the previous upscalers from version 5, they can be found conveniently located under any normal image upscale from `--v 6` and `--niji 6`, and utilizing these features will also require more GPU resources, so it's wise to consider this in your project planning.

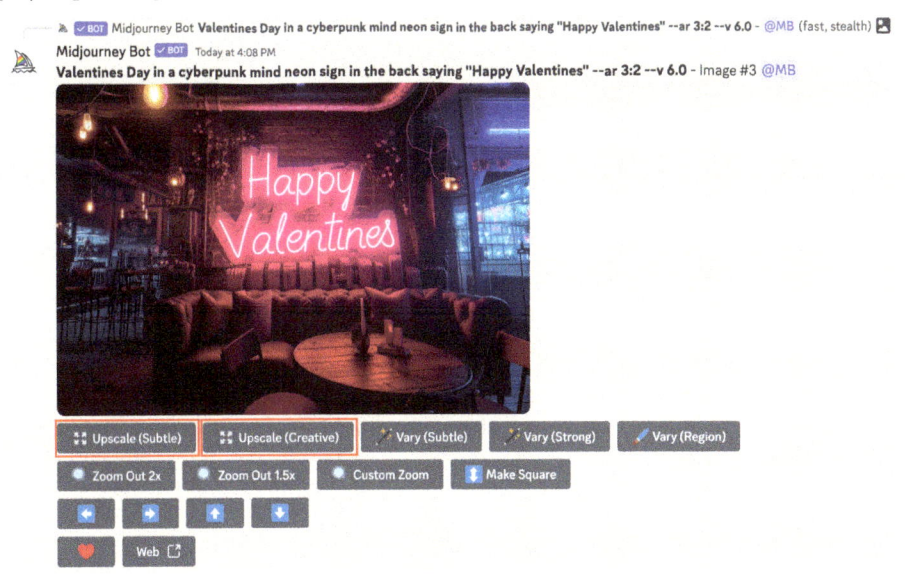

Figure 9.15 – Upscale any image from your grid using any of the U1, U2, U3, or U4 buttons in model version 6 ALPHA or Niji 6, and the two new buttons will appear

266   Unlocking Tips and Tricks and Special Functions

Let's take a closer look at what these enhancements bring to your creative toolkit:

- **Upscale (Subtle)**: Focused clarity

    The **Subtle** feature is crafted to boost the resolution of your images while keeping the original aesthetic intact. It enhances clarity and sharpness, perfect for when you desire a more detailed image without altering its core appearance. This is evident in *Figure 9.16*, where the image went from 1344 x 896 pixels to 2688 x 1792 pixels with no visible changes in the image besides the quality improvement.

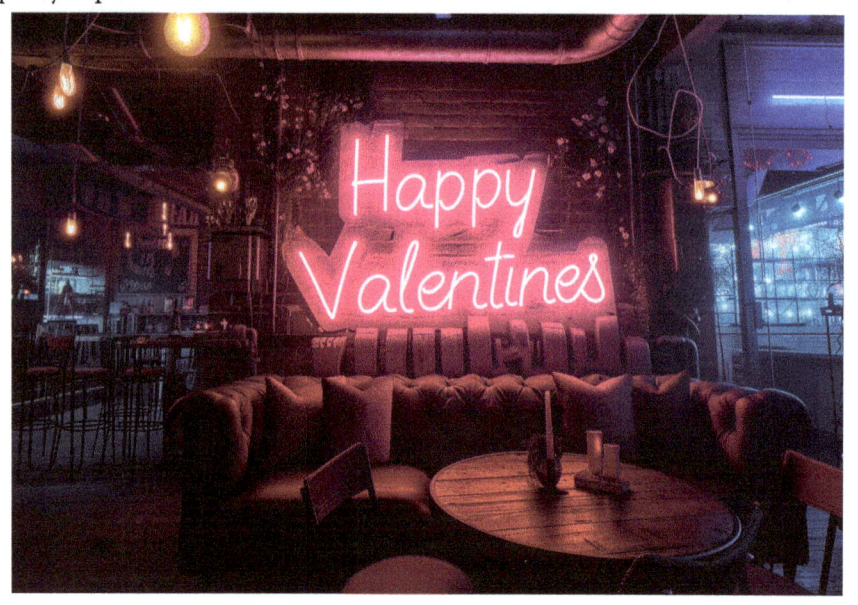

    Figure 9.16 – Quality improvement with Upscale (Subtle), preserving the original image aesthetics

    This approach is ideal for projects where detail and accuracy are key, or an image you want to keep almost untouched with the upscale. It ensures that the essence of your image is preserved, offering a clearer and more defined version of your initial concept.

> **Note**
> **Upscale (Subtle)** is most effective when your main objective is to enhance image quality rather than embark on new artistic explorations. It's suited for scenarios where maintaining the original vision is crucial.

- **Upscale (Creative)**: Creative flair

    The **Creative** feature opts for a bolder path, enhancing resolution while also inviting new artistic interpretations. This feature can modify aspects such as composition and style, adding a creative layer to the upscaling process. As seen in *Figure 9.17*, the image resolution increased from 1344 x 896 pixels to 2688 x 1792 pixels, but with visible changes in the background, objects on the table, and even the pillows and sofa.

Figure 9.17 – Upscale (Creative) at work: Increased resolution (1344 x 896 pixels to 2688 x 1792 pixels) with artistic alterations to background and objects

This approach is ideal for those eager to experiment or inject a fresh perspective into their images, **Upscale (Creative)** can introduce surprising twists and turns to your original work.

But the changes introduced by **Upscale (Creative)** may not suit every project's needs. It's a choice that leans toward experimentation and might lead to results that diverge from your initial expectations.

When deciding on your approach, consider whether your main goal is simply to enhance the image quality. If so, opt for the **Upscale (Subtle)** option, which is designed to improve the image's resolution while preserving its original look and feel.

However, if you're not only seeking quality enhancement but also a dash of creativity, the **Upscale (Creative)** choice might be right for you. This option is ideal for those eager to see how artistic alterations can transform an image, offering a fresh and unique perspective.

Experimenting with both the **Subtle** and **Creative** upscale features on a variety of images will shed light on their distinct effects. This exploration will assist you in determining which upscaling method best suits your ideas.

Midjourney can actually do marvelous work in terms of upscaling, but it's also beneficial to broaden our horizons. There's a whole world of image enhancement beyond what's built into Midjourney, especially when you're aiming for top-notch prints or ultra-fine details. This is where third-party upscaling tools come into play. They're the perfect complement to Midjourney's features, giving us even more power to polish and perfect our images. Let's take a closer look at some of the best external upscalers out there and see how they can take our Midjourney creations to the next level.

Using third-party upscaling tools

While Midjourney's upscaling capabilities are impressive, sometimes you need an extra touch to get your images just right, especially for large-format prints. That's where third-party tools such as **Let's Enhance**, **Topaz Labs**, and **Magnific.ai** come into the picture. These tools can be game-changers in their own right, so let's take a closer look. Let's begin our discovery with Let's Enhance.

Let's Enhance

Start by navigating to **https://letsenhance.io** and create a free account:

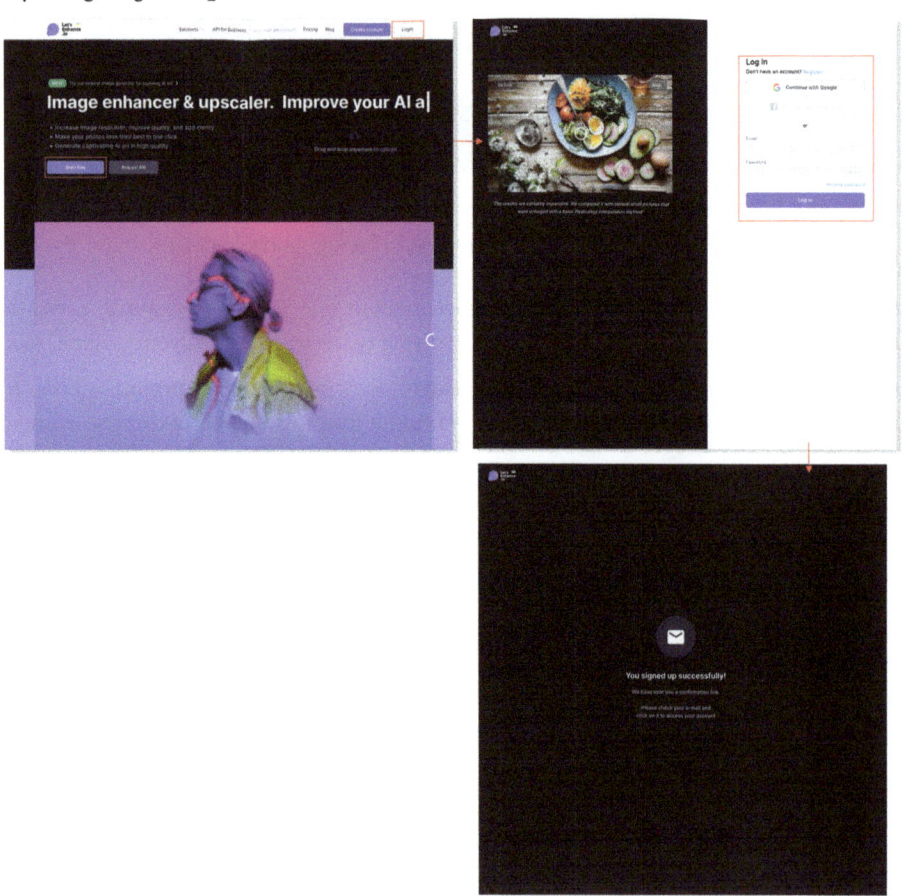

Figure 9.18 – Creating a free account in https://letsenhance.io/

Next, simply upload an image to their website and voilà, you see the transformation. The free plan at Let's Enhance offers 10 credits when you sign up, allowing you to upscale images up to 2 megapixels. This is somewhat similar to what you can achieve with Midjourney's Upscaler 2x, but in Let's Enhance, you have a lot of other features you can use to manipulate the outcome of the upscaler, such as controlling the light, tone, and color intensity before you hit **Start processing** at the end of the page:

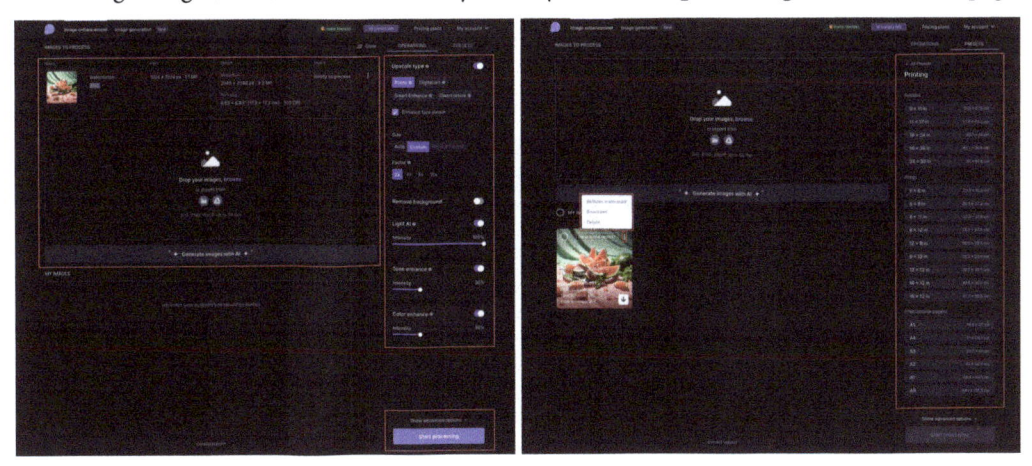

Figure 9.19 – After uploading your image, make the adjustments you want in the right panel

However, there's a catch with the free plan at Let's Enhance – all enhanced images come with a watermark. If you want to remove this watermark and unlock even more powerful features, such as upscaling your images to A0 size (84.1 cm x 118.9 cm) (see *Figure 9.19*), you'll need to upgrade to a paid plan. This upgrade not only frees your images from watermarks but also opens up new possibilities for large-format printing, taking your image quality to the next level.

Figure 9.20 – Upscale 2x from Midjourney (left) versus the 2x upscaler (right) from Let's Enhance

Now, let's shift our focus to another upscaler tool, Topaz Labs.

Topaz Labs

Known for its ability to unearth hidden details and eliminate noise as if it never existed, this software takes image resolution to unprecedented heights. Topaz Labs doesn't stop at images; it's also adept in the art of video upscaling. However, it's important to note that whether for photos or videos, Topaz Labs doesn't offer a free trial, and as it is software, it only supports one kind of subscription, for $199. To know more, go to `https://www.topazlabs.com/`.

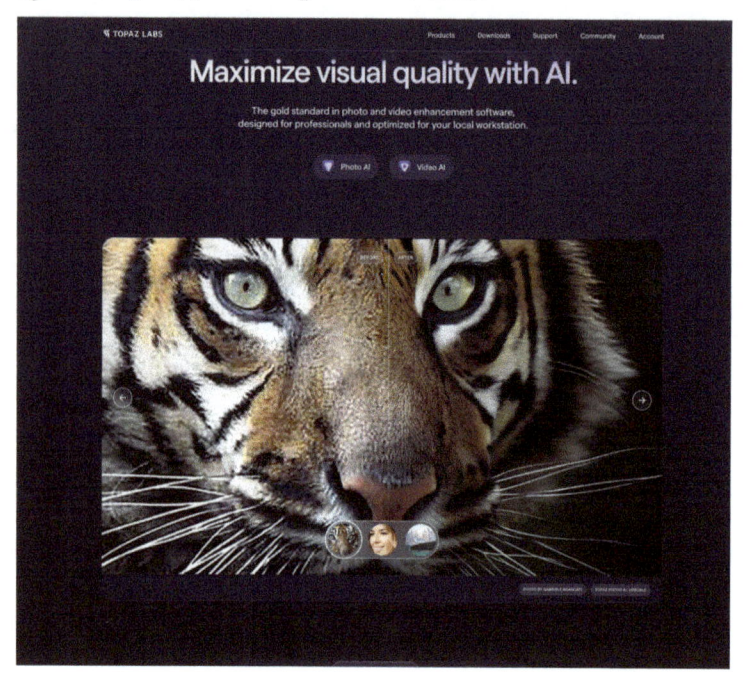

Figure 9.21 – The Topaz Labs website, showing the quality of its enhancement software

Last but certainly not least, we'll take a look at Magnific.ai.

Magnific.ai

I can say that this tool is like having a crystal ball for your images. Besides making the images bigger, it predicts and fills in the details, creating a high-resolution masterpiece from a lower-res starting point. While Magnific.ai, like Topaz Labs, doesn't offer a free plan, it stands out with its more affordable pricing and the option to pay on a monthly basis as it is not a software program, but a web-based tool.

Creating high-resolution imagery for printing   271

Figure 9.22 – On the Magnific.ai website, click on Upscale an image to log in or create an account

Magnific.ai is particularly useful for us, creators working with Midjourney, especially when we're looking to upscale older images created with earlier model versions of Midjourney with less definition, because besides being a powerful image enhancer, Magnific.ai is also very proficient at refining images by removing noise, sharpening details, and enhancing colors. The results are visually striking, making even our old Midjourney images truly stand out. You can check out this tool at `https://magnific.ai/`.

Figure 9.23 – Using the same Midjourney image with Upscale 2x (that I also used in Let's Enhance) on the left, versus the 2x upscale from Maginfic.ai on the right

When picking the right tool for the job in terms of choosing an upscaling tool, it's important to think about what you need. Do you want a quick, easy solution? Or are you after the highest quality, no matter the cost? Maybe you need something that can handle a bunch of images at once. Each of these tools has its strengths, so pick the one that fits your project like a glove. From my experiment, I see that Midjourney produces a very realistic and detailed image, sharp and clear. However, the image is also a bit noisy, and there is some blurriness around the edges. With Let's Enhance, we get a smoother and more polished image than Midjourney with less noise, but still, some parts remain less defined. Overall, Magnific.ai produces the best results. The image is both realistic and detailed, with very little noise.

Figure 9.24 – Comparison of the results of the same image with three different upscalers (2x), from left to right, Midjourney, Let's Enhance, and Magnific.ai

In conclusion, the introduction of Upscale 2x and Upscale 4x marks a significant advancement in Midjourney's digital imaging capabilities. These features not only empower artists and designers with higher resolution outputs but also open up new possibilities for detailed and large-scale prints without the need to look for third-party upscalers. But when you bring in tools such as Topaz Labs and Magnific.ai, you're taking things a little bit further. Whether it's for a stunning print or just to make your digital images pop, these tools are your ticket to image perfection.

Next, we'll dive into essential tips and tricks to refine your prompts. From honing your initial idea to employing strategic emoji use, the following insights will guide you through optimizing your prompts for the most impactful and aligned outcomes.

Tips and tricks to improve your prompts

As we approach the final stages of our journey in understanding the advantages of using Midjourney as a creative tool, let's summarize some key tips and tricks that can be useful for achieving the best prompts and images that closely align with our initial idea. Next, I will create a list of points that can make a difference in any type of prompt and help achieve objectives more effectively:

- **Add descriptors to your prompts**: Grant your prompts another level of sophistication by infusing them with a vibrant palette of styles, descriptors, and techniques. Dive in and see what you can create with these examples:

  - **Photographic terms**:

    - **Bokeh**: Evoke soft, blurred backgrounds for a dreamy, ethereal effect
    - **Film grain**: Imprint authentic film textures, reminiscent of analog photography
    - **Stop motion**: Capture the world in captivating, frame-by-frame animation
    - **Time-lapse**: Depict life's fleeting moments in mesmerizing sequences

  - **Light and flash**:

    - **Mid-morning**: Capture the golden hues of the sun's ascent
    - **Blue hour**: Embrace the tranquility of dawn's transition into day
    - **Monolight**: Employ a single light source for a dramatic, focused effect
    - **Cinematic**: Conjure the ambiance of Hollywood blockbusters with dramatic lighting
    - **Speedlight**: Utilize a handheld flash for dynamic illumination and storytelling

  - **Camera descriptors**:

    - **Lomo**: Embrace the imperfections and nostalgic charm of Lomography
    - **Polaroid**: Capture spontaneous moments with vibrant, instant nostalgia
    - **Pinhole**: Experiment with the unique perspective and aesthetic of pinhole photography
    - **Night vision**: Explore the world under the cloak of darkness with eerie, infrared imagery
    - **Hyperspectral**: Delve into the hidden colors and patterns of the world around us

  - **Film**:

    - **8K**: Render images with exceptional clarity and detail
    - **4K**: Experience a higher resolution than HD, offering enhanced sharpness
    - **HD**: Capture images with high-definition quality, suitable for most applications
    - **70 mm**: Emulate the cinematic grandeur of 70 mm film, known for its cinematic depth
    - **Cinerama**: Immerse viewers in panoramic vistas with the expansive format of Cinerama

- **Focus:**
  - **Deep focus**: Maintain sharp focus throughout the image, emphasizing every detail
  - **Unfocused**: Employ soft focus to blur elements and create a dreamy, artistic effect
  - **Shallow focus**: Selectively isolate subjects, placing them in sharp relief against a blurred background

- **Exposure:**
  - **Short**: Capture fast-moving subjects with crisp details without motion blur
  - **Long**: Render fleeting moments, such as star trails or flowing water, with artistic blur
  - **Double**: Experiment with double exposure techniques to blend multiple images into a single composition

- **Art styles:**
  - **Iconography**: Create symbolic representations with recognizable motifs and archetypes
  - **Diorama**: Construct miniature worlds with intricate detail and a sense of scale
  - **Gothic**: Conjure a sense of awe and mystery with imposing architecture and dramatic lighting
  - **Humorous**: Infuse your artwork with lightheartedness and playful elements
  - **Realistic**: Capture the world around us with precise detail and accuracy
  - **Surrealist**: Challenge conventional perceptions by blending reality and fantasy
  - **Vivid**: Employ bold, saturated colors to create a dynamic and energetic atmosphere
  - **Whirly**: Experiment with swirls, patterns, and organic forms to create visually captivating imagery
  - **Xmaspunk**: Blend festive elements of Christmas with a cyberpunk aesthetic
  - **Kawaii**: Embrace the cuteness, innocence, and playfulness associated with Japanese kawaii culture

- **Colors:**
  - **Aestheticcore**: Explore a range of pastel hues and soft, muted tones
  - **Candycore**: Embrace the vibrancy and sweetness of candy-inspired colors
  - **Color grading**: Manipulate the color palette to create a specific mood or atmosphere, for example: **/imagine prompt** `Futuristic, 3000s, young beautiful model, face skin is partially covered with holographic mirror transparent scales, pink color grading --v 5.2`:

Figure 9.25 – Example of an image using the color grading descriptor in the prompt

- **Pantone**: Utilize specific Pantone color palettes to achieve a defined aesthetic
- **CMYK**: Experiment with the primary colors of cyan, magenta, yellow, and black for diverse color combinations
- **Duotone**: Create a striking visual effect by limiting the image to two colors

- **Avoid banned words**: It's crucial to be aware of Midjourney's content moderation policies and the use of banned words. Midjourney maintains a strict policy against generating images or using prompts that include explicit, disrespectful, or abusive content. This includes avoiding themes related to violence, adult content, gore, and other visually shocking or disturbing elements. To ensure your prompts are compliant and respectful of these guidelines, consider the following tips:

    - **Replace banned words and use synonyms or alternate phrasing**: If your prompt inadvertently includes a banned word, consider replacing it with a similar, allowed word or using a synonym. Alternatively, rephrasing your prompt can effectively convey your idea without triggering the content filters. This approach is particularly useful for navigating around words that might be related to banned content, ensuring that you maintain the essence of your prompt while adhering to Midjourney's guidelines.
    - **Understand the categories of banned words**: Midjourney categorizes banned words into groups such as Gore Content, NSFW/Adult Content, and Other Offensive Content. Familiarizing yourself with these categories can help you avoid using such terms in your prompts.
    - **Adhere to Midjourney's rules of conduct**: By using Midjourney, you agree to its privacy policy and rules. Violating these rules, especially by creating images or using text prompts that are inherently disrespectful, aggressive, or abusive, can lead to a suspension or ban from using the AI tool.

- **Keep content PG-13**: Generally, ensure that your images and prompts are suitable for a PG-13 audience. This means avoiding any content that is sexually explicit, excessively violent, or otherwise inappropriate for younger viewers.

  By following these guidelines, you can navigate Midjourney's content moderation system effectively while ensuring that your creative process remains respectful and compliant with the platform's standards.

  Visit `https://docs.midjourney.com/docs/community-guidelines` to learn more about this.

- **Be patient**: Midjourney is still under development, so it's not always going to generate the perfect image on the first try. Be patient and keep experimenting until you get the results you're looking for.

- **Check image address and format**: When using images in your prompts, ensure that the image address is a direct link to an online image, ending with extensions such as `.png`, `.gif`, `.webp`, `.jpg`, or `.jpeg`. In most browsers, you can obtain the image address by right-clicking or long-pressing on the image and selecting **Copy Image Address**.

- **Describe desired elements**: It's more effective to describe what you want in the image rather than what you don't. For instance, if you're creating an image of a mountain scene and mention `no birds`, your image might still include birds. To ensure specific elements are excluded, use the advanced prompting technique with the `--no` parameter.

- **Detail your vision**: The more specific you are with your prompt, the closer the output will align with your vision. Consider detailing the subject, medium, environment, lighting, color, mood, and composition. However, being vague can also be a strategy to invite variety and unexpected elements.

- **Ensure correct image placement and composition**: Image prompts should be placed at the beginning of your prompt. For the prompt to work, it must include either two images or one image combined with text. *Chapter 5* covers the process in detail.

- **Integrate text directly into your images**: Yes!! It's true: with the latest V6 ALPHA model, text generation is now a reality. Simply enclose your desired text within "quotation marks" in your prompt to seamlessly incorporate it into your images. For a practical example, refer to *Figure 9.15*. Additionally, you have the power to dictate the text's appearance by detailing your font preferences. Mention characteristics such as *script*, *calligraphic*, *handwritten*, *decorative*, *gorgeous*, *stylized*, *thick*, or *thin* to refine your results further.

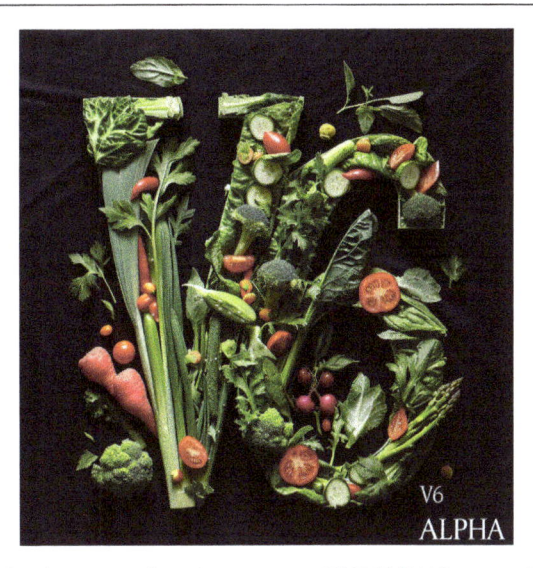

Figure 9.26 – Using a simple prompt /imagine prompt of "V6 ALPHA" text made from veggies --v 6.0

- **Leverage Midjourney's multilingual capabilities**: Midjourney currently supports over 15 languages, including major ones such as English, French, German, Spanish, Portuguese, Italian, Korean, Japanese, Chinese, Russian, Polish, Turkish, and Dutch. However, the quality of results can vary significantly between languages. This variation is due to the differing amounts of training data available for each language, with English having the most extensive dataset. Consequently, prompts in English tend to yield more accurate and consistent results. For users not fluent in English, I recommended writing prompts in your native language and then using translation tools such as Google Translate to convert them into English, or even ChatGPT to help enhance the grammar. This approach can help avoid confusion and improve the quality of the generated images, especially for more complex or ambiguous prompts. In any case, it is always interesting to experiment with prompts in our own language, especially because there are certain concepts that are difficult to translate and that can have surprising results.

- **Optimize your prompt length**: The effectiveness of a prompt in Midjourney is not necessarily tied to its length. While single words or emojis can generate images, they often rely on Midjourney's default style. To achieve a unique look, a more descriptive prompt is beneficial. However, overly long prompts may not always yield better results. Focus on the main concepts and be concise. It's also important to consider that there are slight variations in how prompts are interpreted with each version of the Midjourney model. As each new version is released, the interpretation of prompts evolves and improves. For instance, while this book was being written, the 6.0 ALPHA version of Midjourney was released. Although it is still in the refinement phase, it has been implemented and is available for any user. To access it, you can select **Midjourney Model V6 [ALPHA]** in the settings or add `--v 6` or `--v 6 --style raw` to your prompt. This version is particularly effective with concise, clear prompts and also supports longer prompts.

- **Simplify image prompting**: For mobile users or those seeking a more streamlined process, the **/blend** command can be used to simplify the image prompting process. *Chapter 5* covers the process in detail.

- **Secret way to access Upscale (4x) from V5 in V6 ALPHA**: Upon rendering your image in V6 ALPHA, and if you're working with a square aspect ratio, the initial upscale options (**U1–U4**) present your creation at 1024 x 1024 pixels. At this stage, you're introduced to **Upscale (Subtle)** and **Upscale (Creative)** options, as we have seen before in this chapter, leading to images sized at 2048 x 2048 pixels for Subtle and for Creative. Yet, for those of us seeking the grandeur of a 4x upscale with 4096 x 4096 pixels without resorting to external software such as the ones we've talked about previously, the solution lies within Midjourney itself. Let's see how it can be done in just a few steps with a little trick:

  I. **Start with your original image**: Go back to your pre-upscaled 1024 x 1024 V6 image (see *Figure 9.27*).

  II. **Activate Remix mode**: Ensure the *Remix* mode is active by typing **/remix**, or just activate it in **/settings**. This toggle is crucial for the next steps.

  III. **Employ Vary (Region)**: Choose **Vary (Region)** but instead of selecting an area within your image, draw something *outside* of your image frame using the rectangle marquee tool. This is important because it ensures we don't alter anything within our image, as we're not selecting anything inside the image frame. *This unconventional step is key.*

Figure 9.27 – Choose Vary (Region)

IV. **Switch to V5.2**: In the prompt, change the version from `--v 6.0` to `--v 5.2` and submit it. Since no actual selection has been made, your image remains unchanged:

Figure 9.28 – Select an area outside of your image and change the model version to --v 5.2

V. **Upscale (4x)**: Now, with the new image grid processed as if in V5.2, the **Upscale (4x)** option appears. Click it to apply the magic:

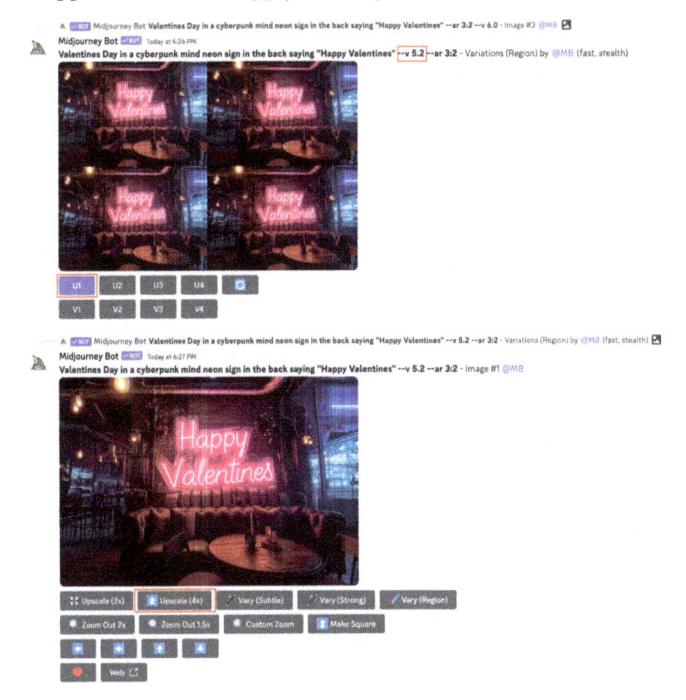

Figure 9.29 – From the new grid, choose an image to upscale using U1, U2, U3, or U4 and the upscalers from --v 5.2 will appear

Another hack for this is instead of using the **Vary (Region)** option, choose **Custom Zoom**, set `--zoom` to `1`, and change the version parameter to `--v 5.2`. A new grid of images will appear. Just upscale any of the images by choosing one of the **U1–U4** buttons and then pressing **Upscale (4x)** (see *Figure 9.29*):

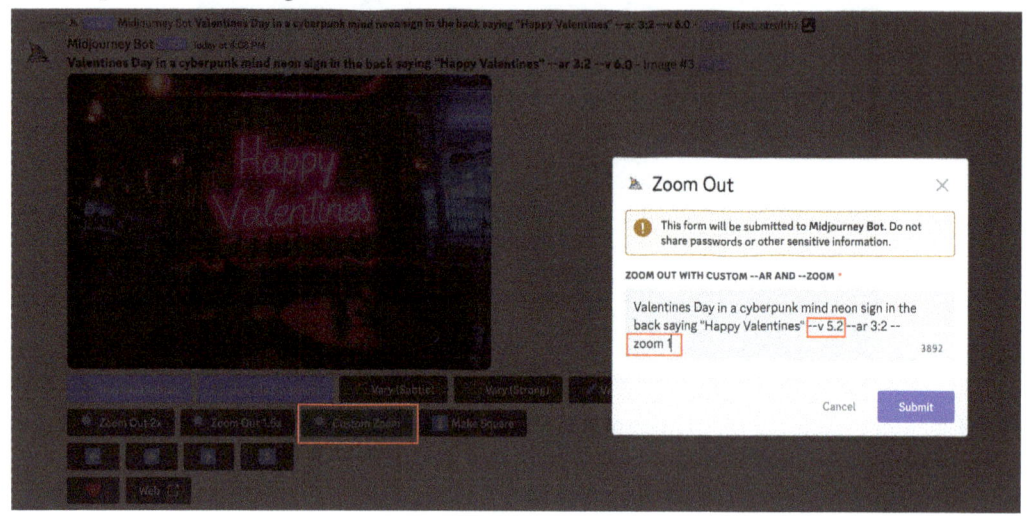

Figure 9.30 – Using Custom Zoom is much quicker to change to --v 5.2 and the results are the same

- **Select the appropriate words and ensure proper grammar**: Midjourney interprets prompts based on keywords rather than grammar or sentence structure. Specific synonyms can be more effective (e.g., using *gigantic* instead of *big*). Simplify your language and avoid unnecessary words. While punctuation such as commas, parentheses, and hyphens can help organize thoughts, remember that Midjourney may not interpret them consistently. Capitalization is also not a factor in Midjourney's interpretation.

- **Tailor images with aspect ratios**: The choice of aspect ratio in Midjourney is more than just a matter of fitting an image into a specific space; it can also alter the outcome of your visual creation. The aspect ratio you select should align with the nature of your subject and the intended use of the image. For example, if you're creating a landscape image intended for a portrait-oriented format, specifying an appropriate aspect ratio such as `--ar 3:4` can significantly influence how Midjourney interprets and renders the scene. This flexibility in defining aspect ratios allows for a more precise and contextually appropriate realization of your creative vision, ensuring that the final image not only fits the desired space but also resonates with the intended narrative or aesthetic.

- **Try different parameters**: Midjourney offers a variety of parameters that you can use to fine-tune the output of your images. Check out *Chapter 4* to know more.

- **Use the power of emojis**: Did you know that Midjourney also excels in the use of emojis in a prompt? These seemingly simple elements, when used strategically, can add a layer of nuance, creativity, and expressiveness to your prompts, leading to even more captivating AI-generated artwork. Emojis, those iconic digital symbols that convey emotions, actions, and objects, can imbue your prompts with a sense of personality and vibrancy. For instance, a prompt that simply reads `A beautiful sunset` might produce a generic image. However, if you add emojis such as 🌅 (sunset), 😊 (hearts), and ✨ (sparkles), you're subtly guiding Midjourney to capture the essence of a breathtaking, magical sunset. You can use emojis to convey the emotional tone you want your image to have, to indicate the actions that you want your characters to take (such as *run*, *swim*, *jump*, or *sing*), or even to identify objects that you want to appear in your images.

Figure 9.31 – Example of the prompt /imagine prompt of sun on the left and the /imagine prompt of 🌞 on the right

> **Emojis for reactions**
>
> In addition to using emojis in your prompts, you can also use emojis to react to Midjourney jobs in Discord. This can be useful for sending images to your direct messages (we have seen before in *Chapter 3*, in the *Beyond /imagine – command list* section, how to get the job id or the seed), for canceling a job in progress, or for deleting an image.
>
> To react to a job, simply click on one of the following emojis:
>
> ✖ – This cancels or deletes the job
>
> ✉ – This sends the image to your direct messages
>
> Note: the ✖ reaction only works with your own jobs.

- **Use collective nouns**: To gain more control over the elements in your image, use specific numbers or collective nouns. For example, *two bunnies* or *a flock of birds* provides more clarity than simply *bunnies* or *birds*.

> **V6: a new world?**
>
> Yes, "V6: a new world?" yes it is!. Imagine: faces so real they breathe, landscapes so vast they echo whispers of forgotten legends, and text woven seamlessly into your visions. This is the power that V6 wields. It ushers in an era of photorealistic wonder, wherein every detail comes alive with stunning fidelity. Yet, remember, this is a world still under construction, a young version, where features might evolve.

I hope these tips help you get the most out of Midjourney!

Summary

Our journey through Midjourney's features has come to a close. Our exploration began with the fundamental concept of facial consistency, with a tool that ensures your characters maintain a cohesive visual identity across generations. This versatility proves indispensable for storytelling and character development, ensuring your creations are consistent and relatable.

Next, we explored the world of upscalers and external applications, discovering their capacity to enhance the resolution of your Midjourney creations. With these tools in hand, you can transform your AI-generated art into breathtaking high-resolution prints, enabling you to share your artistic imagination with the world.

To supplement these technical insights, I compiled a comprehensive collection of practical guidance, encompassing prompt optimization, descriptors, style lists, and even some tips for the newest model version, V6 ALPHA. These practical tips and tricks will equip you to fully utilize the power of your prompts, unlocking a vast expanse of creativity within Midjourney.

This chapter marks the end of our technical exploration, but I know it's just the beginning of your creative adventure. Look forward to the opportunities ahead, and let Midjourney be your ally in bringing to life creations that are both inspiring and impactful.

Happy AI creations!

Figure 9.32 – Happy AI creations (Created with Midjourney by the author)

10
Conclusion: Your Journey with Midjourney

As we draw this guide to a close, we stand at the threshold of a new horizon in digital creativity, thanks to Midjourney. This final chapter is a reflection on the transformative journey we've embarked upon and a forward gaze into the boundless potential of AI in art and design. Through the pages of this book, you've not only navigated the intricate pathways of Midjourney's capabilities but also unlocked a treasure trove of skills that boost your creative expression.

In this concluding chapter, we will summarize the wealth of knowledge and practical skills acquired throughout the book, reinforcing your ability to harness Midjourney's power for your creative projects. We will also gain insight into my vision for the future of Midjourney and AI in general.

In this chapter, we will explore the following topics:

- **Reflecting on the journey and its potential**: We revisit Midjourney's transformative evolution, highlighting its advancements in capabilities and user experience. This section covers the significant updates Midjourney has undergone, the introduction of features such as Style References, and the move toward web-based image generation, making it more nuanced, efficient, and accessible.

- **My point of view and perspectives**: I share my insights into the future of AI in creativity, discussing the potential for even greater integration into our professional and personal lives. We'll explore the existing possibilities that neuro-integration and personalized education hold for the future, and the ethical considerations essential for navigating AI's advancements responsibly.

By the end, you'll not only have a comprehensive understanding of Midjourney's current state but also be equipped to anticipate and navigate its future developments. Let's reflect on the journey undertaken and envision the various ways you can continue to integrate this revolutionary technology into your artistic experiences, pushing the boundaries of what's possible.

Reflecting on the journey and its potential

As we embark on this final chapter, it's essential to pause and consider the transformative path we've traversed together. When starting this book, many of you may have been novices, curious about the burgeoning world of AI and its creative applications, specifically through the lens of Midjourney and its evolution up to the remarkable V6 ALPHA model. Now, as we reach the conclusion, you stand on the precipice of a new creative path, armed with a wealth of knowledge and resources that empower you to craft any image and masterfully control its output.

Reflecting on the evolutionary journey of Midjourney from its inception to the current V6 ALPHA, we witness not just incremental updates but significant leaps in its capabilities and user interface. Since the start of this book, Midjourney has undergone numerous transformations, each version building upon the last, refining its interpretative power and enhancing user experience. The transition from V5 to the ALPHA release of V6 exemplifies this progression, showcasing how foundational learnings are not discarded but improved upon.

One of the most notable evolutions in V6 is the nuanced approach required for crafting prompts, alongside enhanced capabilities for generating text and better interpreting the prompts. These changes underscore a broader trend seen across all versions: Midjourney's increasing sophistication in understanding and interpreting user input. Adapting our prompt-writing strategies has become an essential skill, evolving alongside the tool itself. Each new version introduces a period of adaptation, a demonstration of the tool's growing complexity and our expanding mastery of it.

Looking ahead, we anticipate further updates, potentially seeing versions such as V6.1 or V6.2, and a default V6, each promising new features and improvements. Yet, the core principles and techniques we've learned will continue to serve as a solid foundation for engaging with these future iterations. In the weeks prior to the writing of this book, a new update has come out for this ALPHA model, which allows us to have consistent styles through the **Style References** feature. It works similarly to image prompts (more on this in the *Image prompting* section in *Chapter 5*), but instead of using single images, you provide URLs to multiple reference images that represent the desired style. It's available for both V6 and Niji V6 and it's easy to use. Just follow these steps:

1. After your text prompt, simply add `--sref` followed by the URL of a reference image.
2. Include multiple URLs separated by spaces (e.g., `--sref urlA urlB urlC`).
3. You can assign weights to each reference image using `::` followed by a number (e.g., `--sref urlA::2 urlB::3 urlC::5`). Higher numbers indicate stronger influence (see the *Multi-prompting* section in *Chapter 5* to learn more).
4. Use `--sw` to set the overall strength of the style effect (`0` to `1000`, with `100` being the default).
5. Remember to put your regular text prompts before `--sref`.

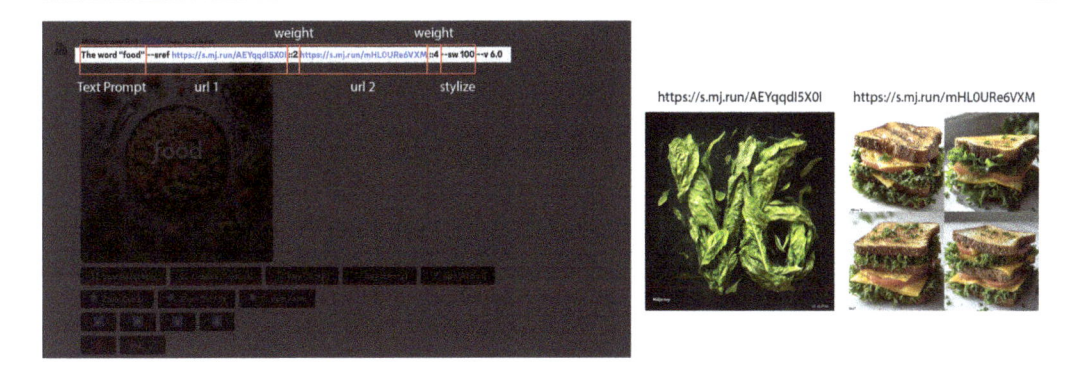

Figure 10.1 – Example of how to use the new Style Reference feature and the images I used as reference

A significant yet still-developing feature, but available to anyone who has generated at least 1,000 images on Midjourney, is the ability to operate Midjourney directly from its website at `https://alpha.midjourney.com`, reducing the reliance on Discord. This transition marks a significant shift in accessibility and user experience. Although not all functionalities available through Discord have yet been integrated into the website interface, this move signals Midjourney's commitment to expanding its user-friendly dimensions. For now, I find myself gravitating toward Discord for its comprehensive control and the ability to fine-tune prompts with advanced parameters that are not all yet available on the website. However, the trajectory is clear: as the website interface becomes more robust, adapting the extensive knowledge and techniques covered in this book for web-based interactions will likely become more straightforward and intuitive.

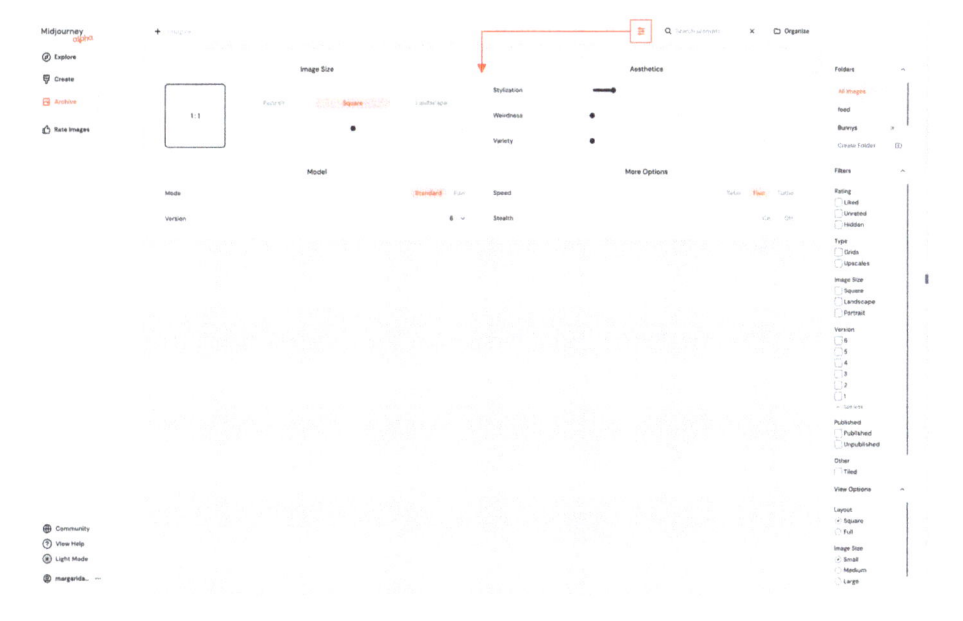

Figure 10.2 – The new site, where you can directly generate your image without the use of Discord

It's fascinating to see how the website, through selection boxes and buttons, offers new ways to manage creativity. Users can organize images into folders, rate them, choose their display mode–whether to view only grids or upscales–and sort them according to preference. The site provides a range of options from image size to the way we wish to view and organize images. The access to the community gallery is the same as in the old site: you can explore projects others have embarked on, discern trends, or often learn new words and terms, serving as inspiration, and all our images are accessible in one place. Additionally, the website interface significantly streamlines the creation process with the introduction of sliders and settings for immediate customization. Directly from the platform, users can select the model version, the type of stylization desired, and even account settings such as choosing between *Relax*, *Fast*, and *Turbo* modes for processing speed. The convenience extends to prompt management as well, with features enabling users to search through their prompts efficiently. This level of direct control and customization, accessible through intuitive sliders and settings, further exemplifies how the website is designed to cater to the diverse needs and preferences of its user base, offering a more tailored and efficient creative experience. This shift to the website interface simplifies the creative process, providing tools and features that foster a more organized workflow.

As we wrap up our look at the latest features and improvements of Midjourney, it's clear that we're just scratching the surface of what's possible with AI in our creative work. This journey has been about learning to use a tool and exploration of how AI can augment our creativity, enhance our marketing strategies, and redefine the boundaries of art itself. To the artists, creatives, and marketing professionals who have joined me on this journey, you've witnessed firsthand the potential of AI to transform ideas into visual realities with unprecedented ease and flexibility.

Next, let's look at the big picture–how AI is not just a tool for making things easier or faster, but a partner that opens up new doors to creativity and innovation. Let's explore what this means for us, the creative community, as we continue to welcome AI and see where it can take our ideas next.

My point of view and perspectives – the next leap, AI's role in our creative and everyday lives

The integration of AI into our creative and technological interactions is broadening the scope of what's possible. AI is revolutionizing the way we create and communicate, becoming deeply connected to our very thoughts and intentions. In the fields of creativity and marketing, AI is no longer a mere buzzword but a key player in transforming ideas into reality, reshaping our digital environments, and redefining the core of creative and marketing roles.

Imagine a world where the ability to create stunning visual content is limited not by one's technical skills or resources but only by the expansiveness of their imagination. This is the reality AI is creating. Picture yourself brainstorming for your next project. Instead of starting from a blank slate, you're collaborating with AI to produce images, videos, and even 3D models that align with your vision. This enhancement of the creative process, facilitated by tools such as Midjourney for imagery, Runway ML for video editing, and platforms such as Cap-Cut, makes creation and editing quicker, more efficient, and limitless.

Figure 10.3 – Working with AI is the present, not the future (Created with Midjourney by the author)

This impact extends beyond individual projects. In the architecture scope, for instance, AI tools such as Midjourney are empowering architects to explore unconventional design concepts. Imagine feeding the program a description such as `a sustainable skyscraper inspired by natural rock formations`. Within moments, Midjourney generates a multitude of initial ideas. This will trigger innovation and allow architects to bridge the communication gap with clients by presenting new concepts early on, fostering better feedback.

The world of visual communication is also undergoing a transformation with AI. Artists and designers can utilize Midjourney to create fantastical creatures and hyper-realistic portraits or establish the mood and atmosphere for a scene. This empowers them to bring characters, scenes, and environments to life for various projects such as movies, video games, and graphic novels. Midjourney excels at creating both fantastical and realistic imagery, offering a valuable tool for concept art, storyboarding, and illustration, as we've seen already in this book.

Advertising is another field where Midjourney is making a significant difference. Using a prompt such as `a photorealistic image of an athlete defying gravity during a basketball dunk` will give you, within minutes, a captivating image generated by Midjourney that could be the centerpiece of a future ad campaign. This efficiency allows brands to experiment with a wider range of creative ideas and identify the ones that resonate most with their target audience.

This new era of creativity inspires us not to think that AI is here to replace our creative process but to enhance the future roles of professionals. How can we adapt to a world where AI tools are our creative collaborators? What new opportunities might this technology bring? It also highlights the importance of ethical considerations. It's crucial that AI serves to augment human creativity and diversity, not diminish it.

As we explore the vast possibilities AI introduces to our creative capabilities, we must also address the challenges. How do we maintain a balance between AI assistance and the essence of human creativity? What measures can we implement to ensure the ethical application of AI in creative fields?

Let's start by chatting about what this balance between AI and human creativity and the enhancement of professional roles means for us folks in creative professions and marketing. Think of having the ability to generate not just static pictures but entire video sequences or even 3D environments from a brief description. This is about amplifying our resources. These tools are giving us the power to go beyond our current capabilities, allowing us to produce more awe-inspiring work without the constraints of our budgets or individual skill sets.

Think about it. A single person can now do what once took a whole team of specialists, thanks to AI automating the grunt work. This isn't just about making things easier or faster; it's about making them possible. We're talking about creating content that stands out, that tells a story in a way we could only dream of before, all while keeping an eye on the budget.

But here's the kicker: as much as this is a game-changer for creating content, it's also a huge win for our work-life balance. Initially, the excitement of exploring this new AI-powered frontier can lead to an overwhelming influx of possibilities, paradoxically leaving us with less free time as we dive into the exploration and mastery of these tools. It's a phase of discovery, learning, and, indeed, adventure, as we navigate through the capabilities and potential of AI in our fields.

Figure 10.4 – It's easy to enter the AI rabbit hole and be overwhelmed by all the knowledge and tools out there (Created with Midjourney by the author)

Yet, as we become more adept at integrating AI into our creative processes, we start to unlock its true value: *the ability to enhance our work-life balance*. By offloading routine tasks to AI, we free up precious time, allowing us to focus on the essence of our creativity and strategic planning, and, most importantly, to reclaim time for our real lives. The narrative shifts from one of exploration to one of empowerment, where AI becomes a tool that not only elevates our professional capabilities but also enhances our personal well-being.

In this new era, the potential for creatives and marketers to deliver content that was previously unthinkable within the constraints of traditional budgets and timelines is monumental. Think of AI as a tool for efficiency and scale, poised to not only revolutionize our professional capabilities but also significantly enhance our personal well-being.

Figure 10.5 – Choose the best AI tools to work with you (Created with Midjourney by the author)

This journey through AI's impact on our creative and everyday lives lays the groundwork for a deeper exploration into the possibilities that lie just beyond the horizon. As we transition from recognizing AI as a powerful collaborator in our current creative processes, we now turn our gaze forward, to the untapped potential of AI in merging our thoughts with digital creation itself.

The evolution of content creation into a neural future

As we witness the exponential growth of text-to-video capabilities, it's clear we're not just taking incremental steps forward; we're leaping. This progression is similar to transitioning from the art of painting portraits to the dynamic world of film directing, where each frame is the power of imagination, now augmented by AI. The impact of this evolution extends far beyond entertainment, reaching into education and personal communication, and setting the stage for profound transformations.

But the innovation journey doesn't halt with video. We're on the cusp of breaking into the next frontier: text-to-3D. Imagine the possibility of typing out a description and watching it come to life as a three-dimensional object or environment that you can interact with through virtual or augmented reality. This leap opens up new avenues for art creation and alters our interaction with the digital world and with each other, blurring the lines between imagination and reality.

At the forefront of these advancements are industry leaders such as Apple, who are pioneering the integration of AI into creative tools. Their research team recently unveiled a novel model called **MLLM-Guided Image Editing** (MGIE) that allows users to modify photos with simple verbal instructions, eliminating the need for traditional photo editing software. While this model has not yet hit the commercial market, it's accessible for exploration on GitHub (`https://github.com/apple/ml-mgie`), and a demo is available on Hugging Face Spaces (`https://huggingface.co/spaces/tsujuifu/ml-mgie`), inviting users to experiment with its capabilities. Moreover, the introduction of technologies such as the *AI Pin* (see more on this at `https://hu.ma.ne/aipin`) is embedding intelligence directly into our physical world, further diminishing the gap between what we can dream and what we can create.

The revolution extends into the domain of education, where AI is redefining the learning experience. Gone are the days of one-size-fits-all education; AI now provides personalized learning journeys, tailored to the pace and style of each individual learner. This approach, akin to having a personal tutor, enhances understanding, reduces frustration, and fosters a deeper love for learning. Let's imagine a platform called *AI Tutor*, which uses smart algorithms to figure out what a student knows, how fast they learn, and what they like to study, from math to art history. Right from the start, it asks questions to understand the student better. As they learn more, the AI adjusts the difficulty and types of questions to keep things challenging but achievable. It's like having a personal tutor available at any time, constantly updating to meet the student's needs.

Take language learning as an example. *LinguaBot AI*, while a fictional concept, shows us what's possible. This AI tutor would offer personalized language lessons, focusing on vocabulary, grammar, and even conversation. It would give feedback right away to help improve areas such as pronunciation and sentence structure, making learning more about understanding and less about memorization.

While *LinguaBot AI* is an idea, real apps such as Duolingo and Babbel are making learning languages with AI a reality today. Duolingo adjusts lessons based on how well you're doing, and Babbel changes its review sessions to better fit your learning style. These apps show how AI can tailor education to our individual needs, leading us toward a future where learning tools are even more personalized and effective.

So, from our fictional *LinguaBot AI* to apps we can use right now, it's clear that AI is changing how we learn. These advancements suggest a future where education is personalized and more engaging and successful for everyone. This AI evolution is promising to erase the lines between digital learning and traditional tutoring, and making personalized education accessible to learners everywhere!

Looking ahead, the potential for AI to become intricately linked with our neural processes presents a tantalizing vision of the future. The concept of thought-driven interaction with digital dimensions, reducing our reliance on physical interfaces, is rapidly transitioning from speculative fiction to tangible reality. In my opinion, this neuro-integration promises a seamless fusion between human intention and digital execution, heralding an era of boundless creativity and interaction.

Figure 10.6 – Maybe in the future, technology and AI will be literally in front of our eyes (Created with Midjourney by the author)

The democratization of creation and innovation through neuro-integration is perhaps its most profound promise. By eliminating the need for specialized technical skills, it opens the door for everyone, including those with disabilities or limited access to traditional education, to express themselves and innovate. This inclusive approach has the potential to set free a wave of creativity and invention from a diverse array of voices and perspectives.

However, as we approach this horizon, we must also navigate the complex ethical landscape that accompanies these advancements. Protecting privacy, ensuring consent, and upholding autonomy are paramount as we integrate AI more deeply into our mental and emotional spaces. Developing neuro-integration technologies within a robust ethical framework is essential to safeguarding individual rights and dignity.

Moreover, the potential for misuse of these powerful technologies calls for stringent safeguards and ethical guidelines. By engaging a broad coalition of stakeholders, including ethicists, policymakers, technologists, and the general public, we can steer the development of AI and neuro-integration toward a future that reflects the best of human values.

As we stand on the brink of these transformative developments, it's clear that the journey of AI integration into our lives and creative processes is becoming increasingly entangled with the essence of human thought and creativity. This journey not only promises to expand the horizons of what's possible but also challenges us to direct these advancements in ways that enhance human well-being and foster a more inclusive, equitable, and imaginative world.

The future of AI and neuro-integration offers a vision of creativity and interaction that bridges the gap between the human mind and the digital world. As we venture toward this future, our focus must remain on leveraging these technologies to benefit humanity, ensuring that they enhance our lives while upholding and reflecting our deepest values and principles.

As we face the advancements in AI that promise to transform content creation and bring us closer to a future where our thoughts can directly shape digital realities, we also stand at a crucial crossroads. This journey, filled with innovation and unlimited possibilities, also compels us to pause and reflect on the direction in which we are heading. The transition from imagining the limitless potential of neuro-integration to considering its implications marks a crucial juncture in our exploration of AI's role in our lives. It's here, at this intersection, that we shift our focus from the technological phenomena on the horizon to the ethical landscape that underpins them.

Ethical landscape and the path forward

Exploring the ethical implications of AI is crucial as we integrate these technologies more deeply into our lives and society. The advancements in AI offer incredible opportunities for growth and innovation but also present challenges that require careful consideration and proactive management.

One of the most pressing concerns in the field of AI is the issue of bias. AI systems learn from vast datasets, and if these datasets contain biases, the AI's decisions will reflect them. This can perpetuate and even exacerbate existing inequalities in society. To counteract this, developers and researchers are focusing on creating more balanced and diverse datasets and developing algorithms that can identify and correct for biases. The goal is to create AI systems that make decisions based on fair, equitable principles, ensuring that the benefits of AI are accessible to all, regardless of background or demographic. But still, the potential for AI to be misused is a significant concern. From deepfakes that can spread misinformation to autonomous weapons systems, the range of harmful applications of AI is wide. It's imperative that ethical guidelines and robust legal frameworks are established to govern the use of AI. This includes international cooperation to set standards that prevent the development and deployment of AI in ways that could harm individuals or society. Education and awareness are also key, ensuring that both creators and users of AI understand the ethical implications of their work and the technologies they use.

Figure 10.7 – Are we teaching AI the right path? (Created with Midjourney by the author)

As AI becomes more adept at generating content that mimics human creativity, distinguishing between what is created by humans and what is produced by AI becomes more challenging. This distinction is crucial for preserving authenticity, copyright, and the value of human creativity. Developing digital watermarking techniques and ethical guidelines for disclosing the use of AI in content creation are steps toward maintaining transparency and trust in the digital landscape.

The aspiration for AI goes beyond replicating human intelligence; it aims to transcend human limitations and biases, offering unbiased judgments and enhancing our ability to identify misinformation. By embedding ethical considerations into the development and deployment of AI systems, we can ensure that these technologies serve as tools for positive change, supporting fairness, integrity, and creativity.

The vision of AI as a beacon of ethical and intellectual advancement is ambitious but attainable. It requires a concerted effort from developers, policymakers, and society as a whole to guide the development of AI in a direction that upholds and advances human values. By setting new standards for ethical AI, we can work toward a future where technology not only enhances our capabilities but also contributes to a more just and equitable society.

Figure 10.8 – We must work to create an AI system that offers unbiased judgments (Created with Midjourney by the author)

The role of AI is not just as a tool or a collaborator but as a catalyst for reimagining what is possible, pushing us toward a future where technology and humanity coexist in harmony, guided by the highest principles of ethics and equity. This path forward is one of optimism and responsibility, embracing the transformative potential of AI while steadfastly committing to shaping a world where technology serves the greater good.

Summary

Picture this: a future where AI, through tools such as Midjourney, fuels your creative fire, effortlessly handles your daily grind, and unlocks new ways to express your thoughts. The line between imagination and reality is blurring. AI tools are putting artistic power into the hands of everyone. Midjourney, as we've explored in this book, stands at the forefront of this revolution, offering a glimpse into how deeply AI can integrate into our creative processes.

With AI, and particularly through our expedition with Midjourney, our connections will deepen, innovation will surge, and we might even gain a deeper understanding of who we are. The platform has already begun reshaping how we approach art, design, and expression, serving as evidence of the transformative power of AI in the creative domain.

While the excitement for what's to come is palpable, it's also crucial to approach these developments with a sense of responsibility and ethical consideration. The power to create, amplified by tools such as Midjourney, is being placed into our hands in ways we've never seen before. How we choose to wield that power, through the images we craft and the messages we share, will shape the future of our physical and digital society. From creation to implementation, we must champion a responsible path for this transformative technology.

By working together, we can navigate the challenges that come with this new era of creative technology, such as job shifts and data privacy, ensuring that AI becomes a force for good. Platforms such as Midjourney are not just tools for artistic exploration but beacons of what's possible when technology and humanity work hand-in-hand to create a brighter tomorrow.

How do you see yourself utilizing AI, and Midjourney in particular, in your creative endeavors, and what potential hurdles do you anticipate?

To answer that, I invite you to stay connected and continue this exploration beyond the pages of this book. For more insights, inspiration, and updates on the evolving landscape of AI in creativity, follow me on LinkedIn (https://www.linkedin.com/in/margaridabarreto/) and Instagram (https://www.instagram.com/margarida.barreto28/). Scan the QR code to access my Linktree, where you'll find a wealth of information, ideas, and resources tailored to fuel your creative endeavors:

Figure 10.9 – Follow me on social media

Moreover, if you're eager to double down on inspiration and engage with a community of like-minded creatives, I encourage you to join the *AI/CC :: Artificial Intelligence Creative Community* on LinkedIn (https://www.linkedin.com/groups/14114059/) and Discord (https://discord.com/invite/aicc). Here, you'll have the opportunity to share your work, learn from others, and experience the collective journey of creativity powered by AI. Let's enjoy the ride together, embracing the boundless possibilities that AI and platforms such as Midjourney unfold for us.

Index

Symbols

/prefer option command 196-198
/prefer suffix command 198-200
--raw and --stylize parameters
  used, for adjusting 206-208
/shorten command 176-179

A

aerial photography 161
AI Tutor 292
ambient lighting 170
anime style 143
artificial intelligence (AI) 3, 4
  concept 5
  content creation, evolution into
      neural future 292-294
  ethical considerations 6
  ethical landscape and path forward 294-296
  LLMs 6
  role, in creative and everyday lives 288-291
artificial intelligence (AI) art
  applying, in generative art 6-8
  future 9
  merging 6

artificial intelligence (AI), categories
  general AI 5
  narrow AI 4
  superintelligent AI 5
artificial intelligence (AI), subfields
  computer vision 4
  machine learning 4
  natural language processing (NLP) 4
art styles
  anime 143
  cartoon 144
  comic book 145
  cyberpunk 156
  diorama 274
  glitch art 155
  gothic 274
  humorous 274
  iconography 274
  ink drawing 146
  kawaii 274
  low poly 154
  manga 143
  oil painting 147
  pastel drawing 148
  pencil drawing 149
  pixel art 152
  pop art 156

Index

realistic 274
Sci-Fi 150
surrealist 274
vector art 153
vivid 274
watercolor painting 150
whirly 274
xmaspunk 274
astrophotography 161

B

back angle shot 174
backlighting 169
basic parameters, Midjourney
   aspect ratios 74-77
   chaos parameter 77-80
   fast parameter 81
   image weight parameter 81
   no parameter 82, 83
   quality parameter 84-86
   relax parameter 86
   repeat parameter 86
   seed parameter 87, 89
   stop parameter 90
   style parameter 90
   stylize parameter 91, 92
   tile parameter 92-94
   turbo parameter 94
   video parameter 94, 95
   weird parameter 96, 97
bird's eye-view shot 174
Blend mode
   using 106-109

C

camera angles 170, 175
   back angle 174
   bird's-eye view 174
   Dutch angle 173
   eye-level 172
   high-angle 171
   low-angle 171
camera descriptors
   hyperspectral 273
   lomo 273
   night vision 273
   pinhole 273
   polaroid 273
cartoon style 144
chaos parameter
   examples 77-80
cinematic lighting 166
cityscape photography 160
colors
   aestheticcore 274
   candycore 274
   CMYK 275
   color grading 274
   duotone 275
   pantone 275
comic book style 145
cyberpunk style 156

D

describe command 134
   benefits 134
   implementation examples 135-140
Discord 17, 18
   exploring 18-21
dots per inch (DPI) 262

D

double exposure photography 163
Dutch angle shot 173

E

evocative style 157
exposure
  double 274
  long 274
  short 274
eye-level shot 172

F

film
  4K 273
  8K 273
  70mm 273
  Cinerama 273
  HD 273
focus
  deep focus 274
  shallow focus 274
  unfocused 274

G

generative adversarial networks (GANs) 7
generative AI (GenAI) 3
generative art 3
glitch art style 155
graphics processing units (GPUs) 195

H

high-angle shot 171
high-resolution imagery
  creating, for printing 262, 263
high-speed photography 162
Hugging Face Spaces
  reference link 292
hyper-realistic visuals 62

I

icons and logos
  brand identity 230-232
  creating 230
  precise control feature, using 233-238
image prompting 109-112
  using 106
image resolution 262
images, with Style Tuner
  adjusting, with --raw and --stylize parameters 206-208
  experimenting, with random styles 208-211
  fine-tuning 200-205
  style codes, combining 205, 206
ink drawing style 146
InsightFace library
  URL 254
InsightFaceSwap Bot 254
  step-by-step guide, to face swapping 254-262
INSwapper feature 257

K

knolling photography 160

L

landscape photography 159
large language models (LLMs) 5
latent space 8

Index

legacy and special parameters, Midjourney 97
  aspect ratios 97
  hd parameter 98
  sameseed 98
  stylize parameter 99
  test models 99
  upscaler 100
  version parameter 101

legacy upscaler parameters
  beta upscale 100
  light upscale 100

Let's Enhance 268, 269
  URL 268

light and flash terms
  blue hour 273
  cinematic 273
  mid-morning 273
  monolight 273
  speedlight 273

lighting
  ambient lighting 170
  backlighting 169
  cinematic lighting 166
  portrait lighting 168
  Unreal Engine lighting 167
  volumetric lighting 165

LinguaBot AI 292
low-angle shot 171
low poly style 154

M

macro photography 161
magical realism style 158
Magnific.ai 268-272
  URL 271

manga style 143
Midjourney 3, 9, 17, 215
  accessing 18
  AI technology 9
  benefits, exploring 12-14
  Discord, connecting to 21-26
  ethical concerns 10-12
  evolutionary journey 49
  journey and potential 286-288
  legacy and special parameters 97
  legal challenges 10-12
  logging into 21-27
  parameters 72, 73
  preferences, customizing 196
  settings, overview 186-196
  version evolution 10

Midjourney Bot 27
  inviting, to server 28, 29
  server, creating 27
  testing 29, 30

Midjourney, evolutionary journey
  Niji versions 64
  V1 and V2 49-54
  V3 54-58
  V4 54-60
  V4, features 59
  V5.0 model 63
  V5.1 63
  V5.2 model 62
  V5 to V5.2 61, 62
  V5, to V5.2 60

Midjourney imagine 36-39
  command list 45-49

Midjourney, preferences
  /prefer option command 196-198
  /prefer suffix command 198-200

Midjourney Prompt Generator
    (MGPM) 180-182
  URL 180
Midjourney prompts 36-45
Midjourney storytelling 223, 224
  example 224-230
Midjourney V6 ALPHA 265
  upscaling options 265-268
MLLM-Guided Image Editing (MGIE) 292
monolithic domain 4
moodboards
  brainstorming partner 222, 223
  brand identities, conceptualizing 221
  business and marketing strategies 218
  event planning 219
  generating 216, 217
  interior design project 220
mood styles 157
  evocative 157
  magical realism 158
  Moody 157
moody style 157
multi-prompting 113-118

N

nature photography 160
negative weights 114
nested permutations 128
neural networks (NNs) 7
Niji models
  versions 64
Niji models, versions
  Niji V5 65-69
  prompts, creating 64
  V4 64, 65

O

oil painting style 147

P

parameters, Midjourney
  basic parameters 72-74
  examples 74
  legacy and special parameters 97
  legacy parameters 72
pastel drawing style 148
pencil drawing style 149
permutations 118
  and parameters 119, 120, 126
  nested permutations 128
  styles and aesthetics 126
  weights and image references,
    testing 130, 131
photogram 164
photographic styles 158
  aerial photography 161
  astrophotography 161
  cityscape photography 160
  double exposure photography 163
  high-speed photography 162
  knolling photography 160
  landscape photography 159
  macro photography 161
  nature photography 160
  photogram 164
  portrait photography 158
  still life photography 159
  street photography 162
  underwater photography 162
  wildlife photography 160

photographic terms
   bokeh 273
   film grain 273
   stop motion 273
   time-lapse 273

photorealistic images 239, 240
   practical applications 241, 245
   professional product mockups, creating with Midjourney 245-252

pixel art style 152
pop art style 156
portrait lighting 168
portrait photography 158
product mockups 238
prompt
   keywords, for effective prompt 175-183
   tips and tricks, to improve 272-282

Q

quality 54

S

Sci-Fi 150
still life photography 159
street photography 162
style codes
   combining 205, 206
Style Raw parameter 62
styles 140
   art styles 143-156
   examples 140-142
   light styles 165-170
   mood styles 157, 158
   photographic styles 158-164

Style Tuner
   technical details 210
stylize 54

T

third-party upscaling tools
   Let's Enhance 268, 269
   Magnific.ai 270-272
   Topaz Labs 270
   using 268
Topaz Labs 268-270
   URL 270

U

underwater photography 162
Unreal Engine lighting 167
Upscale 2x 262-264
Upscale 4x 262-264

V

variational autoencoders (VAEs) 7
vector art style 153
volumetric lighting 165

W

watercolor painting style 150
wildlife photography 160

‹packt›

packtpub.com

Subscribe to our online digital library for full access to over 7,000 books and videos, as well as industry leading tools to help you plan your personal development and advance your career. For more information, please visit our website.

Why subscribe?

- Spend less time learning and more time coding with practical eBooks and Videos from over 4,000 industry professionals
- Improve your learning with Skill Plans built especially for you
- Get a free eBook or video every month
- Fully searchable for easy access to vital information
- Copy and paste, print, and bookmark content

Did you know that Packt offers eBook versions of every book published, with PDF and ePub files available? You can upgrade to the eBook version at packtpub.com and as a print book customer, you are entitled to a discount on the eBook copy. Get in touch with us at customercare@packtpub.com for more details.

At www.packtpub.com, you can also read a collection of free technical articles, sign up for a range of free newsletters, and receive exclusive discounts and offers on Packt books and eBooks.

Other Books You May Enjoy

If you enjoyed this book, you may be interested in these other books by Packt:

OpenAI API Cookbook

Henry Habib

ISBN: 978-1-80512-135-0

- Grasp the fundamentals of the OpenAI API
- Navigate the API's capabilities and limitations of the API
- Set up the OpenAI API with step-by-step instructions, from obtaining your API key to making your first call
- Explore advanced features such as system messages, fine-tuning, and the effects of different parameters
- Integrate the OpenAI API into existing applications and workflows to enhance their functionality with AI
- Design and build applications that fully harness the power of ChatGPT

Unlocking the Secrets of Prompt Engineering

Gilbert Mizrahi

ISBN: 978-1-83508-383-3

- Explore the different types of prompts, their strengths, and weaknesses
- Understand the AI agent's knowledge and mental model
- Enhance your creative writing with AI insights for fiction and poetry
- Develop advanced skills in AI chatbot creation and deployment
- Discover how AI will transform industries such as education, legal, and others
- Integrate LLMs with various tools to boost productivity
- Understand AI ethics and best practices, and navigate limitations effectively
- Experiment and optimize AI techniques for best results

Packt is searching for authors like you

If you're interested in becoming an author for Packt, please visit `authors.packtpub.com` and apply today. We have worked with thousands of developers and tech professionals, just like you, to help them share their insight with the global tech community. You can make a general application, apply for a specific hot topic that we are recruiting an author for, or submit your own idea.

Share Your Thoughts

Now you've finished *The Midjourney Expedition*, we'd love to hear your thoughts! Scan the QR code below to go straight to the Amazon review page for this book and share your feedback or leave a review on the site that you purchased it from.

`https://packt.link/r/1-835-08697-7`

Your review is important to us and the tech community and will help us make sure we're delivering excellent quality content.

Download a free PDF copy of this book

Thanks for purchasing this book!

Do you like to read on the go but are unable to carry your print books everywhere?

Is your eBook purchase not compatible with the device of your choice?

Don't worry, now with every Packt book you get a DRM-free PDF version of that book at no cost.

Read anywhere, any place, on any device. Search, copy, and paste code from your favorite technical books directly into your application.

The perks don't stop there, you can get exclusive access to discounts, newsletters, and great free content in your inbox daily

Follow these simple steps to get the benefits:

1. Scan the QR code or visit the link below

   https://packt.link/free-ebook/9781835086971

2. Submit your proof of purchase
3. That's it! We'll send your free PDF and other benefits to your email directly